Living Well Now and in the Futur

For Linda,

With gratitude
for your good friendship,
mark and

Randy

Living Well Now and in the Future

Why Sustainability Matters

Randall Curren and Ellen Metzger

The MIT Press
Cambridge, Massachusetts
London, England

First MIT Press paperback edition, 2018

© 2017 Massachusetts Institute of Technology

This book was set in Stone Sans and Stone Serif by Toppan Best-set Premedia Limited. Printed and bound in the United States of America.

Library of Congress Cataloging-in-Publication Data

Names: Curren, Randall R., author. | Metzger, Ellen, author.
Title: Living well now and in the future : why sustainability matters / Randall Curren and Ellen Metzger.
Description: Cambridge, MA : The MIT Press, 2017. | Includes bibliographical references and index.
Identifiers: LCCN 2016033654 | ISBN 9780262036009 (hardcover : alk. paper)
ISBN 9780262535137 (paperback)
Subjects: LCSH: Sustainability--Moral and ethical aspects. | Democracy--Moral and ethical aspects. | Justice--Moral and ethical aspects. | Interpersonal relations--Moral and ethical aspects.
Classification: LCC GE196 .C87 2017 | DDC 174/.4--dc23 LC record available at https://lccn.loc.gov/2016033654

10 9 8 7 6 5 4 3 2

For Rayna Oliker, Liesel Schwartz, Richard Feldman, Cameron Schauf, and the students and educational leaders like them everywhere who are creating communities of constructive engagement in advancing sustainability.

Contents

Preface

What is sustainability? Why does it matter? Is the idea of sustainability even coherent? The premise of this book is that sustainability is profoundly important, but little understood. Many people who agree that it is important also say it is hard to define, and although it is acknowledged to be an inherently normative concept there is uncertainty about exactly what its normative content is. The aim of this book is to clarify the nature and normative aspects of sustainability, why it matters, what stands in the way of it, how it might be achievable, and what is required of us to achieve it. This is an unavoidably exploratory and highly synthetic, cross-disciplinary enterprise, a work at the boundaries of philosophy and science that combines them only as well as a philosopher and scientist working together over the past seven years have learned to do. In the interest of facilitating the widest possible understanding of sustainability, we explain as much as possible of the fundamentals.

Sustainability concerns the impact of our present ways of life on opportunities to live well in the future, so this is a book for everyone with an interest in the prospects for humanity—for themselves, their children, and anyone else they care about—in the overburdened world stretching before us. In parenting and teaching, we face the awkward reality of being spokespersons for the world—adults who must believe in the prospects for living well on this planet and in our capacity to equip our children and students to live well without destroying those prospects for others. This is what sustainability is about: We must believe in a world of opportunity but are collectively living in ways that are diminishing opportunity. Under these circumstances, it is best that we understand what we are doing so that we can limit the damage and protect what is most valuable in our lives and the world.

The collaboration from which this book has grown began in early 2009, when one of us (Curren, a philosopher) visited the other (Metzger, a geologist) in San José while completing a manuscript on sustainability and education for a policy pamphlet series in the United Kingdom.[1] In the course of a day on the northern California coast renewing a friendship that has spanned most of our lives, we discovered that the work we were independently engaged in was increasingly focused on sustainability. We were soon educating each other across the disciplinary and geographic divides that separated us and planning this volume, though it has proven to be a much longer journey than we anticipated. Cross-disciplinary collaboration and synthesis are critically important to progress in understanding and advancing sustainability, and our participation in this collaborative endeavor has been uniquely gratifying. A hard truth about sustainability is that it cannot be adequately addressed within the confines of any one discipline. No one existing discipline is adequate to the task, and we are all collectively still groping toward the kind of synthesis that is needed. The language of sustainability, metrics and models of sustainability, field of sustainability studies, and curricula of sustainability are all still being invented. Whatever progress the two of us have made in this larger enterprise, there remains much work to be done. We hope the contents of this volume will prove beneficial to all who read it and useful to those who wish to advance the conversation further.

Randall Curren
Rochester, New York

Ellen Metzger
San José, California

Acknowledgments

Our intellectual and personal debts incurred in the development of this volume are too numerous to fully recount, and we extend our apologies to those we fail to mention by name. For their encouragement, invitations, wise counsel, and comments on draft chapters that were formative in the development of our understanding of sustainability and work on this volume, Curren owes thanks to the following individuals and groups: Hugh LaFollette, Christopher Winch, Christopher Schlottmann, Adam Frank, Bob Smith, Dale Jamieson, Gabriel McMorland, Katrina Korfmacher, Carol Gould, Michael Katz, Ruth Cigman, Elaine Unterhalter, Yvonne Raley, Samuel Gorovitz, Michael Davis, Brian Schrag, Jan Everett, David Burns, Lisa Newton, Michael Bonnett, Douglas Bourn, Ann Finlayson, Joseph Tainter, Karen Berger, Shunzo Majima, Takahiko Nitta, Mark Collier, Susannah Smith, Wade Robison, Evan Selinger, Michael Boylan, William FitzPatrick, Susan Santone, Matt Ferkany, Danielle Allen, Didier Fassin, the Theory2 reading group in the School of Social Science at the Institute for Advanced Study, David Schmidtz, James Garvey, Anthony O'Hear, Kory Schaff, Laura Wray-Lake, Michael Tiboris, Sigal Ben-Porath, Richard Dees, Gina Schouten, Alexandra Streamer; the sustainability research interns, Gena Akers, Daniel Muller, Rayna Oliker, Sarah Studwell, and Shruti Nayar, at the University of Rochester; the Central New York Interface of the Humanities with Science and Technology working group; and colleagues in the Frederick Douglass Institute for African and African American Studies. Curren also thanks audiences at the University of South Florida St. Petersburg, the Rochester Institute of Technology, San Jose State University, the University of London, Hokkaido University, the University of Minnesota Institute for Advanced Study and the University of Minnesota's Morris campus, Marymount University, the New York University Environmental Studies Program, the Royal Institute of Philosophy (London), the University of Rochester, San Diego State University, the University of Colorado Boulder, the University of Pennsylvania, SUNY

Binghamton, Illinois State University, and meetings of the Association for Practical and Professional Ethics, Society for Philosophy and Public Affairs, Philosophy of Education Society, Philosophy of Education Society of Great Britain, Human Development and Capability Association, Geological Society of America, North American Association for Environmental Education, and the American Geophysical Union.

Curren also owes a great debt of thanks to the Institute for Advanced Study in Princeton, New Jersey, and to its School of Social Science for funding him as the Ginny and Robert Loughlin Founders' Circle Member for 2012–2013 and providing an ideal setting for sustained inquiry and writing, as well as to the deans of the College of Arts, Sciences, and Engineering and the provost of the University of Rochester for generously accommodating a research leave for the 2012–2013 academic year. Funding for an earlier phase of his work was provided by the Andrew Mellon Foundation through the Humanities Corridor of Central New York's Interface of the Humanities with Science and Technology working group.

Revised material from Curren's *Education for Sustainable Development: A Philosophical Assessment*, republished online at http://onlinelibrary.wiley .com/doi/10.1111/imp.2009.2009.issue-18/issuetoc, appears with revisions in chapters 1, 2, and 6 with the permission of the publisher and Impact series editors. Revised portions of the following articles by Curren are reprinted by prior arrangement with the editors and publishers of the named publications: "Sustainability in the Education of Professionals," *Journal of Applied Ethics and Philosophy* 2 (September 2010): 21–29 (in chapter 4); "Defining Sustainability Ethics," in *Environmental Ethics*, 2nd ed., ed. Michael Boylan (Oxford: Wiley-Blackwell, 2013), 331–345 (in chapters 1 and 3); "Aristotelian Necessities," *The Good Society* 22, no. 2 (2013): 247–263 (in chapter 3); "A Neo-Aristotelian Account of Education, Justice, and the Human Good," *Theory and Research in Education* 11, no. 3 (2013): 231–249 (in chapter 3); and "Judgment and the Aims of Education," *Social Philosophy & Policy* 31, no. 1 (2014): 36–59 (in chapters 4 and 6).

Metzger is grateful to NASA and the Clarence E. Heller Charitable Foundation for funding in support of San José State University's teacher professional development programs in sustainability and climate change education. She has benefited enormously from conversations with colleagues and collaborators Richard Sedlock, Eugene Cordero, Grinell Smith, Cynthia Schmidt, Susan Santone, and Paul McNamara, members of San José State University's Sustainability Board, and the conveners and participants of workshops presented by the National Science Foundation–funded InTeGrate project (http://serc.carleton.edu/integrate/index.html), which advocates for cross-curricular teaching of geoscience in the context of societal issues. Last but far from least, the teachers of the Bay Area Earth Science Institute and students enrolled in SJSU's Science, Society and Sustainability seminar have contributed their energy, perspectives, and expertise to help launch and guide a daunting but exhilarating and border-crossing journey into a world of new ideas.

We owe special thanks to Jason Blokhuis, Danielle Zwarthoed, and Beth Clevenger and Kathleen Caruso of the MIT Press and to three anonymous referees for their generous editorial guidance and their comments on the August and November 2015 drafts.

Introduction

The aim of this book is to provide an introduction to sustainability that does not simply synthesize diverse ideas about sustainability from a variety of fields, but instead is systematic, probing, and above all *normatively clarifying*. It aims to clarify the nature, forms, and value of sustainability, to survey the problems, catalog the obstacles to achieving sustainability, identify the kinds of efforts needed to overcome those obstacles, and articulate an ethic of sustainability with lessons for governments, organizations, and individuals. Focusing on what is essential to living well, it outlines a theory of justice and just institutions that clarifies the normative core of sustainability: the long-term preservation of opportunities to live well. After discussing the nature of just institutions—and workplaces, schools, and knowledge-producing institutions in particular—the book addresses the growth of complexity in societies and the costs and hazards of competition-driven, sociopolitical complexity. It illustrates many key ideas through a trio of case studies pertaining to energy, water, and food and concludes with a vision and defense of education in sustainability.

Technological innovation is clearly necessary for sustainability, but it is not enough. This book's overarching argument is that unsustainability is primarily a problem of social coordination and that effective and legitimate coordination rests on shared norms of cooperation, common understandings of the nature of sustainability problems, and widely-distributed self-governance as well as government action. The *art of sustainability*—of preserving opportunity to live well into the future—is thus an art of governance involving articulation of and adherence to *fair terms of cooperation* and the creation of institutions that are conducive to people living well both now and in the future, including educational institutions that provide the needed understanding, capabilities, and virtues of cooperation.

Interest in sustainability has grown rapidly, but clarity about the nature, normative aspects, and systemic realities of sustainability has not kept pace.

Interest in sustainability-related matters is evident in the pace of publications concerning climate change, ocean levels, water stress and the ethics and economics of water management, sustainable energy and food, humanity's footprint, environmental degradation, and mass extinctions.[1] It is evident in growing recognition of the damaging impact of human activities on the natural systems on which those activities ultimately rely and in the recent emergence of sustainability science, which crosses disciplinary boundaries to explain the complex interactions between human and planetary systems.[2] It is evident in global, national, and regional climate and sustainability initiatives, from the post-Kyoto climate negotiations to the United Nations' Decade of Education for Sustainable Development. Yet even as acceptance of the importance of sustainability has spread, it has remained a conceptually and normatively murky idea. There could scarcely be a more apt target for scientifically informed philosophical and ethical consideration. The natural scientists who are advancing our understanding of earth systems, social scientists who are advancing our understanding of social systems, policymakers who seek sustainability-enhancing interventions, and public that must be equipped to live well in a rapidly changing world would all benefit from a morally clarifying understanding of sustainability.

Chapter 1, "What Sustainability Is and Why It Matters," aims to bring some clarity to the idea of sustainability—and to do so in a way that supports a coherent and ethically compelling understanding of the normative heart of sustainability. We focus on the essential normative core of a primary *ecological* concept of sustainability and distinguish several other forms of sustainability that are often conflated with it in more amorphous and value-loaded ways of speaking about sustainability. A virtue of this approach is that it provides a basis for detailed inquiry into the ethical core of sustainability and the relationships between several forms of sustainability that people have reason to care about. It is important to the pursuit of sustainability to have a conceptualization of sustainability that both is normatively clarifying and can be "operationalized" or integrated with metrics that can be used to guide policy and measure progress.

Other approaches to defining sustainability are possible, of course, and they have their merits. Definitions can be more or less lexical, more or less stipulative or prescriptive, and may be analytical or intended to clarify the nature of the thing denoted by the term. A comprehensive lexical mapping of the language of sustainability would explain all the ways in which the terms *sustainable* and *sustainability* actually are used.[3] A purely stipulative definition would specify how a newly invented term is to be used, whereas a moderately stipulative or reconstructive definition might prescribe a reformed usage for an existing term. We surmise that the language of sustainability was invented with an action-guiding purpose in mind and offer what we take to be mildly reconstructive definitions designed to clarify the reasons for action. The clarification of reasons for action in pursuit of sustainability flows from our analytical identification of *the long-term preservation of opportunity to live well* as sustainability's defining concern.

Beginning from the premise that sustainability is fundamentally concerned both with the health of ecosystems and with intergenerational justice in human affairs, we distinguish ecological sustainability, throughput (or environmental) sustainability, and sociopolitical sustainability and then define the derivative concepts of compatibility with and conduciveness to sustainability. We identify the long-term preservation of opportunities to live well as the defining form of justice at the heart of sustainability and explain the importance of sustainability through the idea of preserving opportunities to live well and a survey of causes and manifestations of unsustainability that are undermining such opportunities. We note and reject the convention of defining sustainability through the Brundtland definition of *sustainable development* and reject related tendencies to include wider ideals of social justice and democracy as defining aspects of sustainability. This allows us to develop an account of the defining normative aspect of sustainability (a form of *diachronic* justice, or justice across time) without prejudging the relationships between sustainability and *synchronic justice*, or justice with respect to a distribution of relevant goods at a given time. The provision of adequate opportunities to live well in the present is widely perceived to be essential to the pursuit of ecological sustainability and a requirement of justice in its own right, but these matters are better investigated than foreclosed by definitional fiat.

Chapter 2, "Obstacles to Sustainability," identifies a variety of human obstacles to achieving sustainability, and in doing so it provides a deeper understanding of the challenges of sustainability and the means by which these challenges might be overcome. Based on the definition of what is conducive to sustainability introduced in chapter 1, *human attributes, practices, norms, settings, structures, cultures, institutions, systems, and policies* may all be conducive or not conducive to sustainability, and those that are not conducive to sustainability may qualify as significant obstacles to achieving it. We survey obstacles to success in achieving sustainability through uncoordinated individual efforts, conclude that coordinated public and collective efforts are essential, and consider further obstacles standing in the way of such coordinated approaches. The obstacles surveyed include structural determinants of the footprint intensity of consumption; cultural and social factors; systemic failures of public knowledge; aspects of human cognition, emotion, and motivation; and limitations of environmental governance. The character of these obstacles to sustainability demonstrates the need for not only a multiscale and coordinated public response to problems of sustainability, but also self-organized collective efforts to create conditions favorable to global cooperation. We conclude that problems of sustainability are largely problems of social coordination that can only be adequately managed on the basis of fair principles of cooperation. The pursuit of sustainability requires fair terms of cooperation, reflected in norms, practices, and institutions conducive to well-being, because coercive methods of governance have limited efficacy in regulating human conduct and bringing about the changes essential to environmental sustainability.

Chapter 3, "Sustainability Ethics and Justice," identifies the basic elements of a fundamental *ethic of sustainability*, showing how principles of sustainability ethics can be straightforwardly derived from core commitments of common morality to respect others as rationally self-determining persons and to take care not to harm others. We refer to these commitments as belonging to *common morality* for two reasons: First, the endorsement of these commitments by diverse theories of morality and cultural traditions is so strong, widespread, and essential to social cooperation that it is reasonable to regard them as common norms.[4] Second, the common law tradition of England, the United States, and Canada has understood a basic ethic of mutual respect and taking care not to harm others not only as common, or widely shared, but also as so self-evident that it need not be codified in order to serve as a basis for legal liability. The idea that an ethic of sustainability can be derived from these common ethical commitments goes against the grain of widely shared assumptions about the limitations of morality as we know it, yet it simplifies the cultural transformation that sustainability requires if what is required is an implication of principles that are already widely accepted as measures of acceptable conduct and human goodness.

The chapter is an undertaking in normative ethical theory, and a few words about its methodology are in order. Our grounds for accepting the existence of core commitments of common morality are surprising to moral relativists, who assert that ethical beliefs vary widely across cultures and assume on this basis that all ethical beliefs are equally justified. Yet, relativism of this kind is rarely defended by philosophers. It is not clear that there are persistent cultural differences of belief about the basic ethical commitments that concern us, because the differences of belief about specific practices might be explained in a variety of ways. Further, even if there are some persistent cultural differences in fundamental moral belief, it would not follow that the beliefs of different cultures are equally true and their practices are equally justifiable; it would not show that moral realism is false, that moral naturalism is false, or that moral constructivist justifications of morality are faulty. *Moral realism* is the view that there are objective moral truths knowable to human beings. *Moral naturalism* is the view that such knowable moral truths are, or include, truths about what is naturally good or bad for human beings. *Moral constructivism* is an approach to justifying principles of morality and justice that shows why basic aspects of the human condition make it rational to accept these principles as authoritative, whether or not any moral assertions are true or knowable in a strict sense. These are the views that have dominated philosophical inquiry pertaining to the justification of morality in recent decades, and they are all incompatible with moral relativism and subjectivism.[5]

The framework we outline in chapter 3 is a hybrid of naturalism and constructivism, and the claims we make about what is objectively good and bad for human beings are based on decades of empirical studies in psychology. The basic ethic of respect underlying our principles of sustainability ethics can and has been defended on constructivist grounds. We define some cardinal virtues of sustainability ethics, explaining how virtues and principles of sustainability ethics are related to one another. We then frame a conception of politics as an art of sustainability and identify a precedent for it in the earliest classics of normative political theory in Greek antiquity. We provide further background in a section called "Background to a Theory of Justice" on the constructivist political philosophy of John Rawls and its limitations for conceptualizing the preservation of opportunities to live well across periods of time in which the occupational opportunities available in a society change. (The general introduction to Rawls's theory should make it easier for readers who are not versed in political philosophy to understand some aspects of our own approach, which follows the background section, but the critical examination of his theory is less essential.) To better understand what it would mean to preserve for the future opportunities to live well that are as good as those currently available, we outline a theory of justice focused on universally necessary conditions for living well. In doing so, we rely on Rawls's constructivist methodology. We share his commitment to equal basic liberties and rights, democratic institutions, and some form of equal opportunity, but in other respects his theory serves as a foil for our own.

We argue that what people can reasonably expect of their society is that it would seek to enable all its members to live well—to live in ways that are admirable and personally rewarding—and that it would do so with particular attention to bases for living well that individuals cannot provide themselves. Justice would require that the arrangements and institutions of a society facilitate both the *acquisition* of personal qualities essential to living well and the *expression* of those qualities in activities that satisfy basic psychological needs and fulfill related human potentials in ways that are essential to living well. We develop this idea, drawing on the current state of knowledge of what is and is not conducive to people experiencing well-being and life satisfaction—research that is already cited in the literature of sustainability, but rarely discussed in any detail.[6] Our conceptualization of universal necessities for a good life offers a suitably timeless account of what people *need* in order to live well, and so what is institutionally essential to the long-term preservation of opportunities to live well. Crucially, it also implies that continued economic growth is neither essential to nor compatible with enabling everyone to live well. The chapter closes by drawing out some implications of the theory pertaining to fairness in allocating the domestic and global burdens of cooperation in pursuit of sustainability.

Chapter 4, "Complexity and the Structure of Opportunity," uses the theory of justice introduced in chapter 3 to critique three categories of institutions that are basic to personal and collective well-being: epistemic or knowledge-producing institutions, educational or personally formative institutions, and workplaces as institutional settings in which personal qualities are expressed in activity that is more or less characteristic of living well. We address the ways in which these institutions should ideally contribute to creating and sustaining opportunities to live well, and we go on to consider how these institutions actually function in increasingly complex contemporary societies. This extends and deepens the critique of existing institutions that begins in chapter 1 and runs through chapters 2 and 3, and it ends with proposals for reforming the three basic forms of institutions considered to make them more conducive to sustainability: by better advancing problem-driven collaborative research and "full transactional transparency," by focusing less on monetary rewards and material production and more on worker satisfaction, and through reforms that would limit the long-term evolution of the structure of opportunity toward a structure that is ever more ecologically destructive, energy-intensive, and stratified. The chapter incudes a critique of hierarchical integrated systems of education and occupational credentialing and of the way fair opportunity is pursued in the United States more generally. The upshot of this critique is the formulation of a *preservation of opportunity principle* that incorporates a systemically, ethically, and psychologically sound understanding of opportunity and supports cross-generational comparisons of quality and quantity of opportunity (unlike Rawls's theory).

The growth of complexity is a matter of great consequence for sustainability and the preservation of opportunity over time. It alters the structure of opportunity, creating new and sometimes rewarding opportunities, but also greater social and economic stratification and growing energy, material, education, and coordination costs. It thereby challenges conventional thinking about the promotion of equal opportunity and reinforces the view that an adequate conceptualization of sustainability or the long-term preservation of opportunity to live well requires a sound understanding of what is inherently essential to living well. Limits to growth are an inevitable aspect of discussions of sustainability, but the systemic relationships between economic and educational system growth are rarely addressed, even as many call for education in sustainability. We address this omission by discussing the interface of school and work in the context of a revealing model of sociopolitical complexity and collapse, contrasting the US system of *market credentialism* with a more equitable alternative. The resulting account of socioeconomic-educational complexity provides the basis for proposals that might contribute to the long-term preservation of opportunity and reverse what is arguably a broad pattern of declining marginal collective benefit from investments in economic and educational growth.

The themes of chapter 4 reflect our position that ideals of synchronic social justice should not be imported into the definition of sustainability, which pertains to the long-term (diachronic) preservation of opportunity, while adding further detail to our argument that success in the pursuit of environmental sustainability requires institutions and systems that are conducive to preserving the ecological basis of future opportunity, conducive to present fair opportunity, and durable or sociopolitically sustainable. The institutions that structure human activities will need to reconcile current and future opportunities for a long time to come.

Chapter 5, "Managing Complexity: Three Case Studies," identifies problems of sustainability as systemic action problems and presents illustrative case studies in the management of energy, water, and food systems. We have chosen case studies pertaining to energy, water, and food because these are foundational, interacting, and causally connected in a variety of ways with climate destabilization. We begin by examining the widely discussed idea that problems of sustainability are *wicked problems* and conclude that the idea of a *systemic action problem* provides a clearer understanding of the nature of sustainability problems and how to manage them. In doing so, we bring together thematic strands from preceding chapters. The illustrative cases that follow concern the 2010 Gulf of Mexico oil spill, Australia's national water management system, and the changing patterns of food production in the Mekong Region of Southeast Asia. The cases progress in this way, from the local and regional to the national and international, and all are concerned in one way or another with relationships between water, food, and energy systems: the widely discussed *water-food-energy nexus*.

The oil spill case illustrates important ideas about sustainability, complexity, and energy introduced in chapters 1 and 4, together with lessons about governance, the coordination of fragmented expertise, and work. It pertains to the management of risk in a region in which energy and food systems comingle in a body of water. The second case study traces the emergence of a national water management system in Australia, its successes and limitations, and the pressure to dismantle it now that the drought has eased. Its lessons pertain to the benefits of competent authority at the requisite scale, the difficulties of leading diverse stakeholders to agreement and long-term cooperation, the strengths and limitations of market-based management of an irreplaceable necessity of life, and emerging models of collaborative and adaptive cross-sectoral management of water, food, and energy systems. The third case, focused on food and farming in the Mekong Region of Southeast Asia, illustrates many of the same ideas as the second, but with a focus on food production along a river system on which six countries rely. This is a region in rapid flux, in which conflicting visions of new prosperity and traditional ways of life meet in decisions that have immense import for present and future opportunities.

Chapter 6, "An Education in Sustainability," considers how education might equip individuals and societies with understanding, capabilities, and virtues conducive to sustainability. Our purpose here is fundamentally prescriptive. We challenge many aspects of current practices and propose a curriculum that would prepare students to live sustainably. We outline the current state of sustainability instruction and developments in educational policy and practice that define the context for prospective innovations. Various criticisms of UNESCO's Education for Sustainable Development

(ESD) are noted: Is it a well-defined, coherent package? Is it too prescriptive? Does it incorporate an appropriate form of environmental education? Is it grounded in a sound conception of education? These questions are addressed to lay the groundwork for a satisfactory conception of education in sustainability (EiS). Having defined sustainability without reference to sustainable development (SD), we speak simply of EiS instead of ESD, while clarifying the reasons that a sound education in sustainability would include a form of development education (DE) favorable to global cooperation and steps toward global justice essential to such cooperation. We go on to argue that a suitable form of EiS is not only educationally legitimate but also essential to an adequate education.

The argument is grounded in the general accounts of justice, human well-being, and education presented in chapters 3 and 4. The unsustainability of present ways of life is a crucial aspect of the world students must contend with, and we argue that they are *entitled to* an education in sustainability that offers them substantial opportunities to live well in a context of increasing ecological and societal risk. There are reasons of both morality and prudence for global cooperation in advancing sustainability, and we argue that these reasons for cooperating to advance sustainability also make EiS desirable. We propose curricular integration with a focus on the dynamics of interacting systems, problem solving, critical and inventive thinking, and global civic education, all brought together in collaborative project work through which learning can engage real problems. Returning to the themes of chapters 3 and 4, we argue that an education in sustainability will have a greater impact if it is provided through an educational system that offers students better opportunities of their own.

The book's conclusion returns to our general theme that we collectively owe our children opportunities that will enable them to live well without diminishing the opportunities of further generations beyond theirs. Making good on this promise of opportunity will require that we educate them well and reform the institutions, systems, structures, settings, policies, and practices that structure so many of the human activities that have overshot the natural bases on which they rely. We take stock of the book's overarching argument, which leads us not to a specific set of policy recommendations, but to a broad conceptualization of the normative principles that should inform such recommendations.

1 What Sustainability Is and Why It Matters

At its core, the language of sustainability is a way of referring to the long-term dependence of human and nonhuman well-being on the natural world in the face of evidence that human activities are damaging the capacity and diminishing the accumulated beneficial products of the natural systems on which we and other species fundamentally rely. What unsustainability implies is that humanity is collectively living in such a way as to diminish opportunities to live well in the future, and the preservation of opportunities to live well presents itself as the overarching normative focus of concern for sustainability.[1] Expressed in these terms, the opportunities for members of nonhuman species to live well matter, just as the opportunities of human beings matter, and many ethicists and moral theorists argue that the quality of life of sentient nonhumans is morally significant. We find these arguments compelling, but we will confine our attention to human well-being in this book. Understanding the nature and ethics of preserving opportunities for human beings to live well is challenging enough for one book.

Clarifying the nature and value of sustainability is our goal in this chapter, a goal predicated on the conviction that conceptual and moral clarity about sustainability is fundamental to sound decision-making in pursuit of sustainability. We will define a vocabulary of sustainability and identify the relationships between some different forms of sustainability that people have reason to care about. As noted in the introduction, our formulations of terms and definitions will be somewhat reconstructive. Our goal is not to report how sustainability terms actually are used; rather, it is to provide a set of terms that is reasonably precise and analytically clarifying, especially with respect to the normative aspects of sustainability. We locate the importance of sustainability in the preservation of opportunities to live well and convey the urgency of addressing it through a survey of the problems of sustainability that are undermining opportunities to live well in the future.

Sustainability: Conceptual First Steps

The language of sustainability has largely displaced the language of environmental conservation, without committing itself to some defining aspects of the logic of conservation.[2] *Environmental conservation* has signified a responsible and efficient use of natural resources for human benefit, subject to public regulatory oversight and guided by a scientific understanding of resource development and environmental protection. It has long been contrasted with *environmental preservation*, or the designation of wilderness areas, habitats, or species to be protected from human exploitation. Sustainability is concerned with human well-being much as conservation is, but the conceptualization of sustainability has recognized that the dependence of human well-being on the natural environment is richer and more complex than dependence on "natural resources."

Further, the pursuit of sustainability has not been limited to conservation strategies. Natural processes of climate and flood regulation are not "resources" that can be removed from nature and put to some use, for instance, but human well-being depends on them. We now know that these processes are being disrupted by human activities, especially the burning of fossil fuels, so a more encompassing concept than that of natural resources is needed. We also know that the impact of human activities is global in reach, cumulative in effect, and remarkably persistent. The impact of our present activities on global climate and ocean chemistry will persist on the order of ten thousand years, and the persistence of impact on biodiversity will be a couple of orders of magnitude beyond that. Thinking about sustainability has thus developed in connection with new ways of conceptualizing human dependence on nature. The concepts of *natural capital* and an *ecological footprint* are widely used to catalog the forms of such dependence and express the extent of overreliance in terms that are easily understood. The value of ecological footprint analysis (EFA) as a basis for policy is hotly debated, but EFA estimations of human burdens on natural systems are ubiquitous and a useful point of departure.[3]

Beginning from a functional conception of economic capital as "a stock [of some form of assets] that yields a flow of valuable goods or services into the future," Robert Costanza and Herman Daly have defined *natural capital* (NC) as natural assets that yield such flows (or *natural income*) into the future, such as a "population of trees or fish [that] provides a flow or annual yield of new trees or fish."[4] They contrast natural capital with *manufactured capital* (MC) and *human capital* (HC) and distinguish two broad types: *renewable natural capital* (RNC) that is "active and self-maintaining" and *nonrenewable natural capital* (NNC) that is "inactive" or "more passive."[5] Ecosystems are (within limits) self-maintaining, productive units, or RNC. They are described as providing supporting, provisioning, and regulating services, including nutrient cycling and clearing of wastes; soil formation; production of food, fuels, and freshwater; and climate and flood regulation. Their *provisioning services* yield flows of extractable resources or goods: wood, fish, water for drinking, and so on. Fossil hydrocarbons and mineral deposits are passive, producing no services until extracted, making them paradigm cases of NNC. Aquifers vary in the degree to which they are active parts of Earth systems (hydrologic cycles) or passive products of such systems (deposits of fossil water or *paleowater*), but the time scale on which they are replenished is so much shorter than that for fossil hydrocarbons used for fuel that they are sometimes distinguished as *replenishable* natural capital.[6] Costanza and Daly do not directly define *sustainability*, but they identify *total natural capital* (TNC, or the sum of RNC and NNC) as "the key idea in sustainability," and they say in essence that humanity is existing sustainably if it lives within the *natural income*, or the current services being generated by the world's existing natural capital, so that TNC is not diminished.[7]

EFA builds on these ideas. It "accounts for the flows of energy and matter to and from any defined economy [economic throughput from the environment as resources and back to the environment as wastes] and converts these into the corresponding land/water area required from nature to support these flows [i.e., to produce that flow of resources and absorb those wastes]."[8] Working from an estimate of the Earth's *biocapacity*, or capacity to generate useful materials and clear wastes, the ecological footprint (EF) of an economy can be specified as a percentage of the Earth's biocapacity, with an aggregate human footprint above 100 percent representing *unsustainability* or expending more than the natural income generated by the world's existing natural capital. In this way, unsustainability is presented as an environmental analog of outspending total current income by cutting into the productive capital from which current income is generated (RNC, or *principal* in the language of financial investments) or drawing down uninvested savings (NNC). Further analytical detail can be introduced by representing *throughput* or *demand* as a product of population × per capita consumption × footprint intensity (of a unit of consumption), and representing *biocapacity* or *supply* as a product of area × bioproductivity (of a unit of area). *Ecological overshoot* is said to occur when throughput exceeds biocapacity (or demand exceeds supply).

Employing an analysis of this kind, the World Wildlife Fund's *Living Planet Report 2014* calculates human demands on living systems in 2010, the most recent year for which data were available, at over 150 percent of the Earth's renewable biocapacity (what is sustainable).[9] This estimate is a broad measure of systemic risk to human well-being, manifested in the depletion of accumulated products of ecosystem activity and loss of biocapacity. The components of EFA provide more specific information: how far biocapacity is exceeded with respect to cropland, livestock grazing, forest products, fishing, and assimilation of carbon emissions and how much it is diminished in geographic extent by human settlement. The human *carbon footprint*—the area of forest needed to absorb carbon emissions that are not absorbed by the oceans—is currently 53 percent of the aggregate human footprint and the fastest growing component.[10] Consistent with World Wildlife Fund projections, the 2005 Millennium Ecosystem Assessment found that 60 percent of the world's ecosystems and the services they provide are in decline as a result of overuse.[11]

An alternative framework for conceptualizing sustainability was introduced in 2009, based on the idea of *planetary boundaries* (PBs) associated with Earth's biological subsystems or processes and the identification of *threshold levels* of certain key variables that could trigger nonlinear, possibly abrupt, and unacceptable environmental change.[12] Attempts to define an acceptable atmospheric carbon dioxide (CO_2) concentration are often predicated on the idea that a threshold of this kind exists with respect to climate change. The planetary boundaries approach seeks to define thresholds with respect not only to climate but also to the rate of biodiversity loss, rate of nitrogen removal from the atmosphere, quantity of phosphorus flowing into the oceans, stratospheric ozone depletion, ocean acidification, consumption of freshwater, land cover converted to cropland, atmospheric aerosol loading, and chemical pollution. This is presented as the first attempt to "quantify the safe limits outside of which the Earth system cannot continue to function in a stable, Holocene-like state," the *Holocene* being the ten-thousand-year geologic period of highly favorable environmental stability in which human civilizations have existed.[13] The advocates of this approach are frank about the scientific challenges that must be overcome to identify some of the relevant thresholds, but they are confident in asserting that those with respect to climate, biodiversity, and the nitrogen cycle have already been exceeded. With advances in scientific understanding of Earth system processes, this planetary boundary framework is being refined, with updated quantification of most of the PBs and the introduction of a two-tiered approach for several of the boundaries that recognizes regional as well as global boundaries and reflects the importance of cross-scale interactions.[14] Examples of cross-scale interactions might include the impact of global climate change in elevating regional consumption of freshwater or the interaction of climate change and the globally dispersed chemical pollution that is largely responsible for declining regional biodiversity (such as biodiversity loss in coral reef ecosystems, which is mediated by both rising water temperatures and acidification resulting from the uptake of atmospheric carbon).

In contrast to the conservationist idea that natural resources should be efficiently managed, the concept and discourse of sustainability are open to strategies of not only wise stewardship of what humanity uses but also preservation. An accurate accounting of the economic value of ecological services may often justify environmental preservation strategies that shield species, habitats, and regions from direct human use. Preservation may also be defended in connection with sustainability on the grounds that direct experience of the natural world is a vital aspect of human well-being. Whether this is understood in terms of contributions to physical and psychological wellness, aesthetically, spiritually, or in some other way, the value of nature from the point of view of sustainability need not be limited to its value as a form of capital, even if there are reasons to count cultural services as an aspect of natural capital.[15] Some contributions of nature to the quality of human life may indeed depend on regarding nature as having noninstrumental value.[16]

It is also important to note that there are now hundreds of indicators and indices to assess and guide progress toward sustainability.[17] Many pertain to the biophysical state of Earth systems. Others are socioeconomic, such as the Sustainable Development Goals (SDGs; formerly designated the Millennium Development Goals [MDGs]) and measures of well-being or Gross National Happiness.[18] Broadly speaking, the former are concerned with the stability of the natural systems on which human beings and other species rely, whereas the latter are intended to replace or counterbalance economic measures, such as gross domestic product (GDP), that have encouraged unsustainable practices.[19]

Throughput Sustainability vs. Ecological Sustainability

This brief history of usage suggests that the word *sustainable* might be regarded as synonymous with *environmentally sustainable* or *ecologically sustainable*. Wackernagel and Rees describe EFA as providing "an intuitive framework for understanding the *ecological* bottom-line of sustainability," and the idea of an EF is compatible with this, suggesting a fundamental concern with the ecological destructiveness of the imprint human beings leave on the world.[20] A human imprint on the world that destroys RNC or biocapacity is invariably an important aspect of discussions of sustainability, but it is not the only thing at stake; also at stake is whether humanity can sustain the levels of energy and materials flowing through the global economy on the *supply* side. Dependence on dwindling glacial melt water and fossil water extracted from aquifers is a problem of *supply-side throughput sustainability*, and the same is true of dependence on dwindling reserves of fossil hydrocarbons. The use of fossil freshwater has little inherent downstream impact, and if dependence on fossil hydrocarbons had no downstream impact (e.g., if comprehensive sequestration of emissions were in place), then it would be a problem of supply-side sustainability, but not *also* an even worse problem of waste-side sustainability. There is a difference between *supply-side throughput sustainability* and *waste-side throughput sustainability*, in other words. The latter pertains exclusively to the health of ecosystems and preserving their biocapacity; it is a manifestly ecological concept. The former pertains more widely to the flows of energy and matter on which human activity depends.

The upshot is that defining sustainability as living within *total natural income* does not fully align with the ecological aspect of human dependence on nature; it gives independent weight to depletion of NNC, whether or not the unsustainability of supply-side throughput entailed by such depletion yields waste assimilation demands that exceed assimilative biocapacity. The EF is thus not a conceptualization of ecological sustainability or ecological overshoot per se, which would pertain to the stability or biocapacity of ecological systems. Instead, what it captures is throughput sustainability, overshoot, or environmental sustainability, inasmuch as stores of NNC are aspects of the natural environment. What a human footprint above 100 percent signifies is that total economic throughput cannot be sustained indefinitely, because NNC is being depleted without compensating expansion of biocapacity or because takings from biologically active systems (RNC) are cutting into nature's working capital or biocapacity or because waste-side throughput exceeds assimilative capacity or due to some combination of these reasons. Throughput sustainability has a large ecological component, and it is enormously important to the future of human civilization and its constituent societies. It entails ecological sustainability, in the sense that it limits throughput to what ecological systems can continue to produce and assimilate. A civilization that is sustainable with respect to throughput would necessarily be ecologically sustainable, in other words, but an ecologically sustainable civilization is not necessarily one that can sustain its level of economic throughput. A *purely ecological* measure would consider only the waste side of NNC throughput, because it would be concerned strictly with human reliance on and damage to the biocapacity of ecological systems. In contrast to EFA, the planetary boundary framework is a straightforward conceptualization and potential measure of ecological sustainability or the capacity of ecological systems to provide what human beings rely on.

We will begin by defining *ecological sustainability, throughput sustainability*, and some derivative concepts of the same. In the interest of distinguishing these concepts from another form of sustainability that does and should concern many people, we will then define a concept of *sociopolitical sustainability*. We will address the ethics and politics of sustainability separately, in chapter 3. Sustainability is an essentially normative concept, to the extent that a concern about and acceptance of responsibility for the future state of the world and quality of life is essential to it.[21] However, there is a tendency among sustainability theorists to invest the very idea of sustainability, or their definitions of it, with greater ethical specificity than this implies. Rather than treat democratic values or social justice as important *independent* norms of public life and say that sustainability ought to be pursued democratically or that social justice is important, these theorists may incorporate or load democratic constraints on the pursuit of sustainability or requirements of social justice into their definitions of the nature of sustainability itself. Such definitions preempt inquiry concerning the relationships between different forms of sustainability and between sustainability and other things of value. It may be true that if we care about sustainability we should also care about democracy and justice or that just and democratic institutions are among the things we should aim to sustain, but the connections between sustainability, justice, and democracy should not be conceived as *definitional*—except insofar as sustainability is fundamentally concerned with the just preservation of opportunities to live well over time. Our aim is to provide a basis for detailed inquiry into the ethical core of sustainability and the relationships between several forms of sustainability that people have reason to care about.

We understand sustainability as pertaining most fundamentally to the ecological risk associated with living in such a way as to impair the natural systems on which humanity and nonhumans rely, given the unavoidability, breadth, and indefinite duration of reliance on those systems. Sustainability in this primary sense is fundamentally a quality of human practices, the totality of practices of some human collectivity being sustainable if and only if it is compatible with the long-term stability of the natural systems on which the practices rely, *long-term stability* being understood to entail stability or preservation of biocapacity. Long-term stability is compatible with natural variations, but a failure of such stability would imply deviation from the norms of such variation—deviation of a kind that would in time make the totality of practices (the scale and character of the practices) no longer possible. What we mean by the practices relying on the natural systems for which stability is undermined is that the character and scale of the dependence leaves the human collectivity in question without any practical alternatives or feasible substitutes for the natural systems.

The idea of a *human collectivity* enables us to speak of the sustainability of human civilization in the aggregate and also—more easily in a world less interdependent and technologically advanced than our own—the sustainability of a specific civilization, society, or smaller collectivity that relies on an identifiable assemblage of ecosystems. The indeterminacy of what qualifies as long-term stability might need to be resolved for some purposes, but defined in this way sustainability is a coherent and fundamentally important object of concern. On this basis, we offer as the primary meanings of *sustainable* and *unsustainable* the following definitions:

• *Ecological sustainability:* The totality of practices of some human collectivity is ecologically sustainable if and only if it is compatible with the long-term stability of the natural systems on which the practices rely.
• *Ecological unsustainability:* The totality of practices of some human collectivity is ecologically unsustainable if and only if it is *not* compatible with the long-term stability of the natural systems on which the practices rely.

The reference in these definitions to a *totality of practices* requires some explanation. By a *practice*, we mean a structured, norm-governed form of activity making up part of a culture. Practices shape activities; they vary widely in their complexity and the learned dispositions, abilities, and understandings they involve, but their structures and norms give current activities a kind of *momentum* or *forward trajectory*. By a totality of practices and its compatibility with the long-term stability of natural systems, we mean all of the activities of a human collectivity during some time period, considered both with respect to their dependence and impact on natural systems and with respect to the momentum or trajectory of the activities entailed by their structure and norms as practices.

This formulation allows the preceding definitions to capture two senses in which the way of life of a human collectivity would qualify as ecologically unsustainable. On the one hand, a collectivity's total current activities might already be ecologically overextended by using ecosystem services beyond their capacity and detrimentally altering or impairing the natural systems on which the activities rely. On the other hand, a collectivity might have a culture that includes practices with a momentum or trajectory that puts its activities on an unsustainable path or a path toward being ecologically overextended. In either case, we could reasonably describe the collectivity's way of life as unsustainable and conclude that a change of practices is needed to avert calamity, notwithstanding the fact that a civilization is at greater risk if its activities already exceed what ecosystems can support without being degraded. As formulated previously, then, the definition of ecological sustainability classifies a human collectivity as sustainable just in case it is neither currently ecologically overextended nor on a path toward being ecologically overextended, given the structure and norms of its own practices. The definition of ecological unsustainability classifies a human collectivity as unsustainable if it is either currently ecologically overextended or on a path toward being ecologically overextended.

Throughput or *environmental sustainability* and *unsustainability* may be defined similarly, with reference to the provisioning capacity of natural systems:

• *Throughput sustainability:* The totality of practices of some human collectivity is environmentally sustainable if and only if the material throughput on which it relies is compatible with the projected provisioning capacity of natural systems.
• *Throughput unsustainability:* The totality of practices of some human collectivity is environmentally unsustainable if and only if the material throughput on which it relies is *not* compatible with the projected provisioning capacity of natural systems.

As with the definitions of ecological sustainability and unsustainability, the wording of these definitions (*totality of practices* and *compatible with*) allow them to capture two importantly different senses in which a collectivity's way of life might be unsustainable. A civilization is overextended or in a condition of overshoot just in case its activities are currently consuming more throughput than natural systems can provide in the future, but a civilization that is not yet overextended in this sense may be unsustainable in the sense that it operates in such a way as to be on a path toward overshoot. It is important to note that what natural systems *can* provide in the future— their "projected" provisioning capacity—is in part a function of whether and how much their productive capacity is undermined by human practices that are ecologically unsustainable.

Some explanation of the notion of following a "path" toward being overextended may be helpful. We have said that the structure and norms of practices give current activities a kind of momentum or forward trajectory. Consider how norms of respectability and status shape the environmental impact of our clothing, homes, hygiene, transportation, recreation, telephones—or anything else perceived as a measure of status or cause for shame or loss of face if one lacks it. What is the environmental impact of consumption induced by norms of changing fashions in clothing and other personal furnishings or by the further differentiation of such items by function or activity? Consider, similarly, the complex constellation of practices related to preparation for an adult role in society, courtship, procreation, and retirement and whether the structure and norms of these practices stimulate growing consumption that is otherwise unnecessary. Do the dynamics of competition for social and economic position invite ever-rising expenditures and consumption? How do reliance on population growth and rising future consumption manifest themselves in real estate values, investment income, debt management, and the possibility of retirement?[22] In a simple agrarian society, security in old age may be predicated on each individual having a large number of children, whereas a wealthy society may be capable of an adequate rate of savings for everyone's retirement but choose to rely on a growing economy to fill a savings gap. Both approaches exhibit the structure and trajectory of unsustainable *pyramid schemes*, which rely on flows of resources from a continually expanding base and collapse when the expansion can no longer continue.[23] When such schemes are systemic, as they seem to be in the increasingly integrated global civilization of the present age, then the recruitment of an expanding human base may continue apace so that the limitations of the planet's natural systems become the limiting factor—the base of the pyramid that *cannot* expand.

It should be noted in this connection that although economic expansion may contribute to environmental protection in some limited respects, the overall environmental impact of human activity is closely linked to the growth of the world economy.[24] The patterns of economic growth have largely determined the path we are on:

The human presence is now so large that all we have to do to destroy the planet's climate and ecosystems and leave a ruined planet to our children and grandchildren is to keep doing exactly what we are doing today, with no growth in the human population or the world economy. Just continue to release greenhouse gases at current rates, just continue to impoverish ecosystems and release toxic chemicals at current rates, and the world in the latter part of this century won't be fit to live in. But ... human activities are not holding at current levels—they are accelerating, dramatically. It took all of history to build the $7 trillion world economy of 1950; today economic activity grows by that amount every decade. At current rates of growth, the world economy will double in size in less than two decades.[25]

Regarding the *Kuznets curve*, or the observation that the pollution directly affecting populations tends to decline as the populations become more affluent, "there is no consensus on the drivers of changes in emissions," and there is no consensus that other sustainability-related phenomena follow the same pattern as pollution that directly affects affluent populations.[26]

Derivatives of Ecological and Throughput Sustainability

What it means for something to be sustainable is less well-defined when what is said to be sustainable without qualification is a business, technology, mode of farming, or something of the sort. A family farm might be nearly self-sufficient and *operate sustainably* under our definitions of ecological and throughput sustainability, but it would still rely on a stable climate and probably other shared and potentially vulnerable ecological services as well, such as pollinators. Considered as part of a larger human collectivity that is overshooting the capacity of the systems on which it relies, it would not be sustainable. Furthermore, its farming methods may only be sustainable on a limited scale and within a larger human context. In general, a particular practice, technology, system, or institution may exist in a form and on a scale that does not undermine the natural capital (RNC or TNC) on which it or any other human collectivity relies, and to that extent it may be said to be sustainable. Within the context of a particular civilization, some practices may be compatible with ecological or throughput sustainability however widely and frequently they are practiced (perhaps thinking or walking to work), whereas the sustainability of other practices (such as eating meat or driving a car to work) may be subject to limitations of scale, frequency, or both. Only the former could be said to be sustainable without qualification, even in the derivative sense of being compatible with ecological or throughput sustainability.

Systems may be ecologically unsustainable in a related derivative sense if the activities they involve cannot occur at the scale and frequency *required by the system* without causing the society as a whole to be ecologically unsustainable. The present US transportation system is ecologically unsustainable because it is predicated on the idea that most people will rely heavily on personal vehicles that consume fuels derived from petroleum, and the pace of greenhouse gas emissions this entails is incompatible with climate stability and favorable ocean chemistry. The present US transportation system is similarly unsustainable with respect to throughput, because the pace of fuel consumption entailed by the role of personal vehicles far exceeds the rate at which petroleum is being generated by the Earth, let alone within the regions of US jurisdiction. Subject to the qualifications thus noted, a venture's compatibility with both ecological and throughput sustainability is undeniably important in deciding how to proceed.

Similarly, it may often be normatively salient to consider what is *conducive to* ecological or throughput sustainability, and this can be defined as another derivative of the basic forms of sustainability defined previously:

• *Sustainability conducive:* A human trait, attribute, practice, norm, setting, structure, culture, institution, system, or policy is conducive to sustainability if and only if it functions, within its sphere of operation in the existing state of the world, to preserve or promote ecological or throughput sustainability.[27]

What distinguishes *conduciveness to* sustainability from *compatibility with* sustainability is that the latter pertains to activities, or what we might think of as first-order activities, whereas the former pertains to what shapes and regulates those activities: the structures and norms of practices, social systems and settings (such as urban landscapes), and associated forms of regulatory or second-order activity, from informal modes of interpersonal instruction and guidance to institutionalized reward structures, government regulatory efforts, and the offices and occupations they entail. Virtues of character and norms of professional integrity are among the human attributes and norms of practice that may qualify as conducive to sustainability.

Often, what we may have the information to judge, with some effort, is that one course of action is *more compatible* with sustainability or that one regulatory structure or setting is *more conducive* to sustainability than another. We may not be able to quantify the entire value of ecosystem services at stake in a decision or project the full systemic effects (both natural and social) of a decision, but we may nevertheless be in a position to make useful comparative judgments of alternative courses of action or alternative regulatory structures.[28] There may be good reason to say that preserving a wetland is more compatible with sustainability than developing it, or good reason to say that urban settings favorable to low storm water runoff and living near work are more conducive to sustainability than ones that are not.

Sociopolitical Sustainability

A fact glossed over by much of the discourse of sustainability is that institutions, social systems, and sociopolitical regimes may be unsustainable and subject to collapse owing to factors unrelated, or only weakly related, to any tendency to deplete NNC or undermine the natural systems on which they rely. An institution may be compatible with sustainability in the primary ecological sense and also compatible with throughput sustainability yet be *socially* unsustainable, or it might be unsustainable in some combination of these respects. Unregulated free markets may belong in the category of multiply unsustainable systems. They may be unsustainable not only in the sense that they are not compatible with the ecological or throughput sustainability of any human collectivity without suitable regulation; they may also be unsustainable in the sense that, quite apart from any limitations of natural capital, they exhibit a tendency to destroy themselves through acquisitions that ultimately yield oligopolistic markets.[29]

The ecological and environmental sustainability of a civilization has profound ethical importance, but matters are less clear regarding the social sustainability of particular practices, institutions, and systems. We have good reason to prefer practices, institutions, and systems that are conducive to environmental sustainability (which entails ecological sustainability, as we noted), and consequently good reason to prefer that practices, institutions, and systems that are not conducive to environmental sustainability fade away or be reformed. Moreover, the pursuit of environmental sustainability should be predicated on institutions and systems that are not only conducive to environmental sustainability and human well-being but also durable or sociopolitically sustainable. We will need them to work for a long time without placing ever-larger demands on the natural systems that make life on our planet possible.

Furthermore, there is no denying that we depend on the cultures and institutions we have and that their decline and collapse often entail a world of loss and suffering. If ecological and throughput unsustainability are not overcome, they will almost certainly lead to both ecological collapse and sociopolitical collapse. *Ecological collapse* denotes a rapid decline of ecosystem capacity (such as the 90 percent decline in ocean fish populations since the advent of deep-sea fishing in the 1950s) or the disintegration of an ecosystem. *Sociopolitical collapse* may be defined as a "rapid, significant loss of an established level of socio-complexity," manifested in rapid loss of socioeconomic stratification, occupational specialization, social order and coordination, economic activity, investment in the cultural achievements of a civilization, circulation of information, and the territorial extent of sociopolitical integration.[30] Sociopolitical collapse entails grave risk to most of what people value. When sociopolitical collapse is triggered by problems of environmental unsustainability, rapid population loss and displacement are also predictable.[31]

From an ecological and resource perspective, a growing human population may be regarded as a driver of unsustainability, but from an *ethical* point of view, a growing human population is, among other things, a growing number of people subject to increasing risk of suffering grave harm.[32] Stabilizing population, reducing carbon emissions, and taking corresponding steps to respect other planetary boundaries would reduce the risks associated with ecological and sociopolitical collapse. Simplification of ways of life may be essential, but it is not as simple as making do with less opulence. It entails reversing the trend toward ever more specialized and stratified occupational roles and reversing the declining marginal return and rising costs—energy, information, compensation, and regulatory costs—associated with that trend.[33]

Preserving Opportunity vs. Equalizing Opportunity: The Brundtland Definition of Sustainable Development

We said at the beginning of this chapter that the preservation of opportunities to live well is the normative heart of the idea of sustainability. The language of sustainability is a way to express the idea that it is wrong, unfair, or unjust to reject choice-worthy ways of life in favor of ones that undermine the natural basis of opportunity for others to live well in the future. Understood in this way, the idea of sustainability is irreducibly normative, but its normative focus is narrower than is often supposed. Its distinctive focus is on living in a way consistent with an acceptable future, a future in which the natural world is not altered by human activity in any way that would cause the opportunities to live well to be worse *in the future* than they are today (diachronically). Although concern for fair opportunity also favors equal opportunities for everyone within a country or across the world *now* (synchronically), sustainability's defining focus is diachronic, not synchronic. Neither environmental nor sociopolitical sustainability entails synchronic domestic or global equality of opportunity as a defining condition.

This is at odds with a widespread tendency to treat sustainability as a comprehensive ideal of social justice and to treat a well-known definition of *sustainable development* as a definition of sustainability: "development that meets the needs of the present without compromising the ability of future generations to meet their own needs."[34] Defined in this way, the concept of sustainable development entails a standard of global justice, according to which the needs of everyone in the world must be met through economic development. Yet sustainability (a quality) and sustainable development (development that has certain qualities) are not the same thing.

The origins of the term *sustainable development* can be traced to the 1972 UN Conference on the Human Environment convened in Stockholm, during which developing countries argued that progress in environmental protection requires progress in reducing poverty and that economic development to alleviate poverty should not be subordinated to environmental concerns.[35] The outcome was a doctrine of "environment and development," now known as sustainable development, and a Stockholm Declaration of twenty-six principles, including a "fundamental right to freedom, dignity and adequate conditions of life, in an environment of quality" and a universal "responsibility to protect and improve the environment for present and future generations."[36] Under the doctrine of environment and development, governments agreed that development and environmental protection are mutually reinforcing and that countries of the Global North would cover at least some of the cost of environmental initiatives in the Global South through additional development assistance. The agenda of sustainable development was articulated further by the 1987 Brundtland Report, *Our Common Future*; the 1992 Rio Earth Summit (United Nations Conference on Environment and Development [UNCED]); and the 2002 Johannesburg World Summit on Sustainable Development.

The doctrine of sustainable development—that development and environmental protection are mutually reinforcing—is problematic, inasmuch as environmental damage generally increases as economic activity increases. There are identifiable ways in which *targeted* development can reduce environmental damage by introducing cleaner technologies, facilitating better governance of environmental commons, and reducing birth rates by providing women with educational and economic opportunities, but the size of the global economy remains a good predictor of aggregate environmental impact. It is true that as people become wealthier they give a higher priority to local and regional environmental protection, but they also adopt lifestyles that are far more environmentally damaging due to the quantities of energy and materials required to maintain them.[37] What the doctrine ultimately reflects is the political reality that global environmental problems can be fully solved only through global treaties predicated on *fair terms of cooperation*.

In the case of climate disruption caused overwhelmingly by the North's huge head start in economic development based on fossil fuels, it is hard to imagine any basis of cooperation that would not involve major concessions to the vulnerability and needs of the global poor. No climate treaty will be adopted that is not perceived as reasonably fair to signatory nations, nor will such a treaty be implemented through domestic enabling legislation and widely obeyed unless it is widely perceived as fair within signatory nations. Our view of the matter is that justice has moral importance, independently of sustainability, and that it also has strategic importance in the domestic and global pursuit of sustainability. Acknowledging this makes more sense than importing extraneous ideals of justice into the very idea of sustainability.

Major Causes and Manifestations of Unsustainability

Even now, at a time of conspicuously strange weather and water use restrictions in growing parts of the United States, many aspects of the emerging global sustainability crisis are not widely understood or discussed. This is not surprising, because our everyday experience, especially in the prosperous enclaves of the North, provides little insight into the larger patterns and consequences of a rapidly growing world economy. Yet with the growth of technology and economies since World War II, the scale and destructiveness of human activity are now far beyond anything in the past.[38]

As documented by multiple indicators of change in the planetary system, human activity is now a potent biophysical force that significantly influences Earth's land, water, air, and ecosystems on a global scale. Moreover, human-induced changes are accelerating: Since around 1950, the world has experienced greater and more rapid change than at any time in the past twelve thousand years.[39] Writing in the inaugural issue of *The Anthropocene Review*, Anthony Barnosky and his coauthors describe five key negative trends—climate disruption, extinctions, loss of non-human-dominated ecosystems, pollution, and population growth—that have the potential to undermine the ecological systems on which society depends, cautioning that if we remain on the current path it is extremely likely that by the time today's children achieve middle age, "Earth's life support systems, critical for human prosperity and existence, will be irretrievably damaged by the magnitude, global extent, and combination of these human-caused environmental stressors, *unless we take concrete, immediate actions to ensure a sustainable, high-quality future.*"[40] The World Wildlife Fund's *Living Planet Report* (LPR), which has been published every two years since the late 1990s, summarizes its science-based assessments of planetary health using three indicators: the ecological footprint discussed previously, the Living Planet Index (LPI), a measure of population trends for thousands of vertebrate species, and a water footprint that tracks humanity's direct and indirect demands on diminishing freshwater resources.[41] The 2014 LPR and the 2015 *Living Blue Planet Report* (LBPR) paint a troubling picture, which, as noted by World Wildlife Fund International Director General Marco Lambertini, is "not for the faint-hearted."[42] Some key conclusions of the LBPR include the following:

• For more than four decades, humanity's demand has exceeded Earth's biocapacity.
• Between 1970 and 2010, the LPI showed a 52 percent decline in global populations of vertebrate species; in other words, over the course of less than two human generations, these populations have diminished to half of their size.
• The water footprint is already leading to water scarcity in a number of the world's major river basins, and it is estimated that by 2030 nearly 50 percent of the global population will be living in areas of high water stress. Because water is essential to the production of food and energy, water scarcity is already impacting access to food and energy; this impact is likely to grow with increasing population and affluence.[43]

Walt Reid, the 2005 Millennium Ecosystem Assessment's lead author, noted similarly in interviews that local and regional ecosystem collapses are already occurring and that humans are "putting such strain on the natural functions of Earth that the ability of the planet's ecosystems to sustain future generations can no longer be taken for granted."[44]

We offer a brief survey of the leading causes and manifestations of unsustainability, organized around basic aspects of EFA and the three principal forms of sustainability we have identified: ecological sustainability, throughput sustainability, and sociopolitical sustainability. Recall that EFA represents demands on the environment as a product of population × per capita consumption × footprint intensity (of a unit of consumption), and it represents biocapacity or supply as a product of area × bioproductivity (of a unit of area). Overshoot grows with demand and loss of biocapacity owing to reductions of productive area and impairment of bioproductivity. We will first examine demands, delaying consideration of NNC depletion so as to address ecological sustainability first. Within the discussion of demands, we will address sequentially the components of the footprint: built-up land, forests, cropland and grazing, fishing, and carbon.

Rising Demand: Population Growth and Per Capita Throughput

The human population doubled to two billion between 1800 and 1950, tripled between 1950 and 2000 from two billion to over six billion, and is currently rising by more than two hundred thousand people per day. The global average fertility rate is 2.8 live births per woman and declining, but it is not expected to reach replacement fertility (2.06) until 2050. It is often assumed on this basis that the human population will level off around mid-century, but the youthfulness of today's population will yield continued population growth until 2100, unless pandemics, widespread drought, or something else intervenes.[45] The global human population, currently 7.3 billion, is projected to reach 9.6 billion by midcentury.[46] One implication of a human footprint that is already 150 percent of what is sustainable is that a population of 9.6 billion would severely overshoot and degrade Earth's ecosystem capacities if it attempted to sustain per capita economic through-put at or near current levels.

Average per capita energy use meanwhile has increased more than ten-fold since 1800, from less than half a metric ton of oil equivalent per year to about five, and individual water use has nearly quadrupled in that time.[47] Global meat production and consumption increased fivefold between 1960 and 2010, requiring several times the energy and water inputs for a unit of food energy and protein than dietary alternatives. A consequence closely associated with the energy intensity of meat production is that livestock currently account for about 18 percent of anthropogenic (human-caused) greenhouse gas (GHG) emissions, including close to 40 percent of all meth-ane and 65 percent of nitrous oxide.[48] Global automobile production increased sixfold between 1950 and 2010 to over fifty million per year, and there are currently approximately 670 million automobiles in use, or 950 million cars and trucks.[49] The use of these personal vehicles has substan-tially displaced reliance on more fuel-efficient public transportation for many of the world's affluent, and it accounts for much of the world's con-sumption of about eighty-six million barrels of oil each day. Global fish catch and consumption increased fivefold from 1950 to 2005, with the advent of large-scale deep-ocean fishing.[50]

Pollution, an aspect of throughput, is a quantity of something in a place or system that cannot assimilate it fully or without being damaged or altered in a way that reduces its value, functionality, or bioproductivity. Pollution impairs the health of humans and nonhumans and impairs ecosystem productivity in a variety of ways. It is estimated that particulate air pollution alone globally causes an annual loss of 49.7 million healthy years of human life.[51] Other major problems caused by pollution are climate disruption (also known as global warming, climate change, or climate destabilization), nutrient loading of water systems by livestock waste, sewage, and fertilizer runoff (which causes coastal eutrophication and hypoxia or deoxygenated "dead" zones), acid rain, ocean acidification resulting from uptake of carbon from fossil fuel emissions, and stratospheric ozone depletion (which allows dangerous ultraviolet radiation to reach Earth's surface—a different problem from global warming, except that the chemicals that cause ozone depletion are also GHGs).[52]

Impact on Biocapacity: Loss of Biologically Productive Land and Loss of Bioproductivity

Human building over of land removes it from ecological service. Urban sprawl replaces biologically productive forests, pasture, cropland, and wetlands. Highways, dams, and other aspects of built environments act as barriers to the free movement of wildlife, fragmenting habitats and ecosystems and impairing the bioproductivity of what remains of the systems. Engineered landscapes may also significantly alter water cycles and regional climate by removing surface water, reducing transpiration, and replacing permeable surfaces that absorb rainfall with impermeable surfaces (contributing to flooding and reducing water availability).

Loss of bioproductivity is occurring across other components of the footprint as well. Global deforestation has continued at a rate of about thirteen million hectares annually since 2000, causing habitat loss and impairing climate and flood regulation, water availability, and other ecological provisioning services. Deforestation and forest degradation currently account for as much as 20 percent of human-caused carbon emissions.[53] *Desertification* refers to processes that persistently degrade the bioproductivity of land, such as soil loss and compaction, salinization, and loss of vegetation. Drylands, which cover 40 percent of Earth's land surface and are home to over two billion people, suffer from ongoing desertification and declining bioproductivity, estimated to entail an annual loss of total photosynthetic activity of about 2 percent.

Much of this is occurring as a consequence of population pressures and food scarcity that has led to converting pasture or rangelands (lands used for livestock grazing) into marginally productive croplands. In the United States, human-induced dryland degradation is occurring gradually across most of the western half of the forty-eight contiguous states.[54] Overfishing has caused a steep decline in fish populations, which has progressed from the most accessible streams and rivers to coastal areas and finally to the open seas. The populations of large fish species have declined 90 percent since 1950, commercially valuable species including the bluefin tuna are near extinction, and some studies suggest that oceans may be essentially emptied of fish by midcentury.[55] An annual 0.7 percent loss of coastal wet-lands (from development, inundation, and pollution) and extensive damage to coral reefs (from commercial exploitation and from acidification and rising ocean temperatures owing to carbon emissions and climate change) are further impairing ocean ecosystems.[56]

Unprecedented and growing carbon emissions from the burning of fossil fuels, forest loss and degradation, livestock, and other human causes have made climate disruption not simply the most serious problem resulting from pollution, but *the most serious problem facing humanity*. Atmospheric carbon dioxide concentrations have risen from the preindustrial level of 275 ppm to above 400 ppm in 2016 and are now higher than at any time in the last three million years or longer.[57] The pace of emissions has also been accelerating: Global carbon dioxide emissions increased 49 percent between 1990 and 2010 and 3.1 percent annually between 2000 and 2010, reaching annual emissions of ten gigatons (billion metric tons) of carbon in 2010.[58] How sensitive is Earth's climate system to these rising atmospheric concentrations? As observed changes have outpaced the worst-case projections of early climate models, it has become clear that a stabilization target of 350 ppm (50 ppm *below* the 2016 level) would be more compatible with avoiding catastrophe than the target of 450 ppm that was long considered safe.[59] Stabilization at 450 ppm would require reductions of global carbon emissions of about two-thirds (and US emissions reductions of about 90 percent) by 2050 and global carbon neutrality by 2100.[60] It is not surprising, in these circumstances, that large-scale removal of carbon from the atmosphere is contemplated as one part of a comprehensive response to the hazards of climate destabilization.

The causes and consequences of climate change are widely documented matters of scientific consensus. As reported by NASA, "ninety-seven percent of climate scientists agree that climate-warming trends over the past century are very likely due to human activities," and the majority of the world's leading scientific organizations endorse this position.[61] The *Fifth Assessment Report* (AR5) of the United Nation's Intergovernmental Panel on Climate Change (IPCC) is itself the work of four thousand authors and based on thirty thousand climate studies.[62] It strengthens the basic conclusions reported in earlier assessments, though in contrast to the IPCC's *Fourth Assessment Report* published in 2007, AR5 is more cautious about making detailed forecasts (e.g., the date by which glaciers will have disappeared) and more concerned with actions that can be taken to increase societal resilience in the face of growing risk.[63]

This massive *Fifth Assessment Report* was released in three installments between September 2013 and April 2014, and a Synthesis Report was issued in November 2014. An official presentation prepared to accompany the Synthesis Report summarizes some key findings:

• Humans are changing the climate, and it is extremely likely that we are the dominant cause of warming since the mid-twentieth century.
• Temperatures continue to rise, and each of the past three decades has been successively warmer than the preceding decades since 1850.
• Growth in greenhouse emissions between 2000 and 2010 was greater than in the previous three decades.
• Impacts are already underway on all continents and in the ocean, affecting rich and poor countries alike and extending from the tropics to the poles.
• Some of the changes in extreme weather and climate events observed since about 1950 have been linked to human influence.
• Continued emissions of greenhouse gases will cause further warming and changes in the climate system. During the twenty-first century, oceans will continue to warm, global mean sea level will continue to rise, Arctic sea ice will very likely continue to shrink and thin, and global volumes of glacial ice will decrease.[64]

In remarks made at a November 2014 press conference in Copenhagen that launched the IPCC AR5 Synthesis Report, UN Secretary General Ban Ki-Moon offered a blunt assessment of the report's findings: "With this new Synthesis Report, science has spoken yet again with much more clarity and greater certainty. Citizens are increasingly restive—but also eager to seize the opportunities of building a sustainable future. ... Science has spoken. There is no ambiguity in their message. Leaders must act. Time is not on our side."[65]

Drawing on data and findings reported in AR5, the World Bank warned similarly in a 2013 report, *Turn Down the Heat: Why a 4°C Warmer World Must Be Avoided*, that if the existing emissions reduction commitments and pledges made by governments are not fully honored, a 4°C rise may occur as early as the 2060s, making the 2010 heat wave that caused fifty-five thousand deaths and 25 percent agricultural losses in Russia and the 2012 drought effecting 80 percent of agriculture in the United States the new summer *norms*.[66] "There is no certainty that adaptation to a 4°C world is possible," write the report's authors. "A 4°C world is likely to be one in which communities, cities and countries would experience severe disruptions, damage, and dislocation ... The projected 4°C warming simply must not be allowed to occur."[67]

Together, these various threats to biocapacity are responsible for this being an era of mass extinctions, one of very few in Earth's history. About one-quarter of land mammals, one-third of freshwater fishes, 50 to 75 percent of amphibians, reptiles, and insects, and 70 percent of all plants are presently at risk of extinction. As many as three hundred thousand species have become extinct since 1950, and the majority of the ten million or so remaining may disappear in our lifetimes.[68] Biodiversity is concentrated in tropical rain forests and coral reefs, which are among the most endangered ecosystems.

Throughput Sustainability

The causes and manifestations of declining ecological capacity outlined in the previous sections constitute the most pervasive and potentially permanent threats to human well-being. Ecological sustainability is throughput sustainability's dominant component, the other component being the sustainability of reliance on NNC. The NNC on which human societies presently are most critically dependent are fossil water and fossil fuels. Water and energy are critically important in themselves, and both play critical roles in food production. The continuing losses of biocapacity outlined previously put the quantity and quality of global food production at risk, given its obvious dependence on such ecological services as climate and flood regulation, soil regeneration, pollination, and the regenerative capacity of fish and animal populations. The dependence of farm yields on fertilizer, pesticides, herbicides, and fuel—derived overwhelmingly from oil and natural gas—entails further risk.

Fossil water, in the form of glaciers and aquifers, is being depleted and will be under increasing threat from climate disruption. This will contribute to water availability being the most difficult leading edge of climate change.

We referred to the depletion of the Ogallala Aquifer in the US Great Plains earlier in this chapter in note 6, and much the same could be said for many other aquifers in India, China, and elsewhere.[69] The importance of glacial melt water is illustrated by the fact that all of Asia's ten major rivers are substantially and critically fed by melt water from the retreating glaciers of the Greater Himalayas. Water scarcity and disputes almost certainly will intensify as global warming progresses and as local control of water sources collides with commodification and international trade in water.

Fossil fuels are the other most critical form of NNC being depleted. Without petroleum, "modern economies would not exist," but the fossil fuel age will run its course within the life expectancy of children born this decade.[70] US conventional oil production peaked in 1971, as predicted by Hubbert cycle analysis, and it has been in steep and irreversible decline ever since. Based on the same kind of analysis, the US Army predicted that global oil production would peak in 2005 and exhibit the same pattern of decline. That prediction was accurate, inasmuch as 2005 was the year in which the highest rate of production of *conventional oil* was recorded.[71] *Nonconventional oil production*, such as from tar sands, cannot replace the shortfall for long, in part because the *energy return on energy invested* (EROEI) for nonconventional oil production is not favorable. Joseph Tainter and Tadeusz Patzek write: "In the United States in the 1940s, petroleum exploration and production gave an EROEI of 100 to 1. Today ... it averages 15 to 1. ... Already for oil from the Canadian tar sands the net energy can be as low as 1.5 to 1 ... [but] we need net energy of at least 5 to 1 to power a modern complex society."[72] Public debate about piping oil from the Canadian tar sands overland to refineries in Texas has been largely silent on the fact that an energy extraction technology that consumes two-thirds as much energy as it produces is no replacement for one that consumes one-fifteenth, let alone a mere 1 percent, as much. Technological advances extend the search for oil and gas into ever-deeper water and increasingly inhospitable environments and tap unconventional sources with low energy yield and high potential for environmental damage.[73]

The so-called shale gas revolution, driven by high gas prices and advances in the processes of horizontal drilling and hydraulic fracturing, allow production of natural gas from previously inaccessible reservoirs, and the abundance of shale gas has been widely touted as an "energy game changer" that offers a cleaner alternative to burning coal. Some have argued that shale gas will supply a "bridge" to a low-carbon future and eliminate American's dependence on foreign oil.[74] Writing in *Nature*, J. David Hughes has called for a "reality check" on these "exuberant forecasts," explaining that oil and gas from shale are "neither cheap or inexhaustible" and noting that 70 percent of American shale gas comes from fields in which production is flat or in decline.[75] As a counterpoint to claims that shale gas offers energy security and a climate-friendlier alternative to coal while the United States transitions toward a low-carbon economy, many argue that the shale gas boom brings unacceptable environmental risks, including potential depletion and irreversible contamination of groundwater, and prolongs US dependence on fossil fuels, delaying a transition to renewable sources of energy.[76]

The process of shifting from a fossil fuel economy to one based on other energy sources will be long and energy-intensive, and its scale will dwarf the energy transitions of the past. Before petroleum refining and drilling began in the 1850s, the oil used for heat and light in Western Europe and North America was supplied primarily by whaling. The whaling industry was itself on the verge of collapse around 1800 and survived another half-century only through more energy-intensive means—namely, steam-powered factory ships and explosive harpoons. Those innovations made it profitable to hunt whale populations that were sparser and harder to find, but at the cost of declining energy yields (EROEIs). With a global population now about six times what it was during the transition from whale oil to petroleum and a way of life that is six or seven times more energy-intensive, the energy transition now in progress will be anything but routine.

At the present time, there are no ready alternatives to fossil fuels that can be scaled up at a feasible cost to meet present demand, let alone growing demand. The prospects for currently available alternative energy sources were detailed by David MacKay in 2009 and summarized by Tainter and Patzek in 2012.[77] "Most sources of renewable energy have less energy density and a lower EROEI than do most fossil fuels," with the consequence that they will require higher energy inputs per unit of energy made available and a larger share of Earth's surface to produce.[78] Average per capita energy use in the United States is about 250 kilowatt hours per day. Tainter and Patzek estimate that devoting 425,000 square kilometers of the windiest regions to wind power production might yield about forty-two kilowatt hours per day per capita.[79] Covering 350,000 of the sunniest square kilometers with solar panels or concentrators might produce the 250 kilowatts per day per capita now used, but the output would be highly variable.[80] Although wind and solar generating capacity have grown much more rapidly than anyone could have anticipated in 2012, problems of transmission, storage, and maintenance remain.[81] The energy yields on widely promoted biofuels may be negative, and even if they are positive the quantity of vegetation that would be needed to replace a small fraction of the fuels now consumed would be prohibitive.[82] The lifecycle energy yields on nuclear power are also relatively poor, and a scaled-up nuclear energy economy would quickly exhaust the fuel available for use in commercial technologies.[83] The cooling requirements of nuclear power stations also make them vulnerable to interruptions of service during heat waves.

Sociopolitical Sustainability

The basic question of sociopolitical sustainability concerns which features of sociopolitical systems are conducive to their continuing viability and capacity to provide their members with opportunities to live well. What traits of sociopolitical systems are conducive to preserving opportunity and avoiding collapse, or a "rapid, significant loss of an established level of socio-complexity," and all the difficulties and uncertainties that entails?[84] We will pursue this question in the following chapters; we will close this chapter simply by identifying three broad categories of hazards societies face.

The first, and most closely related to ecological sustainability and throughput sustainability, is the risk a society runs in damaging or impairing the natural capital, or biocapacity and nonrenewable resources, on which it fundamentally relies. Analogous to impairing or destroying the biocapacity of RNC is damaging NNC or rendering it unusable, such as by polluting fossil groundwater. A further environmental threat to sociopolitical stability and sustainability is catastrophic storms and coastal inundation by rising sea levels.[85] Bjørn Lomborg and others argue that human societies have become much richer and will likely continue to do so and that the history of commodities in market economies shows that they become more plentiful and cheaper over time.[86] It may be, as Cass Sunstein says, that within the two-hundred-year time scale that Lomborg and others have in mind, "most generations are richer and more informed than those that preceded them," measured by income, consumer goods, and a variety of welfare indicators.[87] However, this scarcely shows that worries about environmental degradation, resource scarcity, and associated risk to human well-being are inconsistent with "the history of the human race" and no threat to sociopolitical sustainability.[88] The historical and archeological records suggest that countless civilizations have flourished longer than ours and collapsed, owing at least in part to environmental and resource problems.[89]

The second hazard for a society is in being structured so as to require *unsustainable flows* of material and energy, so that regardless of any damage the society's activities may cause to the natural basis on which it relies, there will inevitably come a time when those flows will not be adequate and decline or collapse becomes unavoidable. We have referred in passing to unsustainable pyramid schemes, and these are paradigmatic of such structures; dependence on ever-expanding flows of resources limits their viability.

The third hazard consists of a sociopolitical regime being structured in such a way as to progressively undermine its own *legitimacy*, which is to say its ability to justify itself to its members and induce their continued respect for its norms and cooperation in performing its vital functions. This may occur if the rules of economic life allow transactions that divest people of the means to provide for themselves as the wealth of others grows without limit—an evil characteristic of oligarchies, as Plato observed 2,400 years ago.[90] A *legitimacy crisis* occurs when a large proportion of a society's members cease to see the terms on which it requires cooperation as fair or tolerable and find it preferable to resist or defect.

Conclusion

Our aim for this chapter has been to clarify the nature, forms, and importance of sustainability and provide an overview of the major causes and manifestations of unsustainability. Constructive engagement with problems of sustainability also requires that we understand the obstacles to success in overcoming the problems. If unsustainability is an aspect of human activities, it is essential that we understand the variety of human factors that explain why these activities are collectively increasingly unsustainable. Surveying these factors is the task of chapter 2.

2 Obstacles to Sustainability

The sciences of sustainability and economics of ecological cost accounting have advanced far enough to justify the conclusion that declining ecological capacity and climate disruption will cause increasingly severe economic and noneconomic losses, more or less in perpetuity, unless the material throughput of human activities is substantially reduced.[1] We saw in chapter 1 that to prevent these losses, carbon emissions must be reduced by two-thirds by 2050 and essentially eliminated by 2100, and total throughput must be reduced by at least one-third from its 2008 level. Greater reductions might be required for natural and sociopolitical systems to have much *resilience*, or capacity to respond to unexpected challenges. These are aggregate targets, and we have seen that sustainability pertains to the totality of practices of human collectivities, whose prospects are now strongly linked through one global atmosphere and one global economy. A consequence of this linkage is that we now, as one global collectivity, face an unprecedented problem of global coordination.

There are many obstacles that must be overcome in order to solve this problem—obstacles that pertain to the character of human systems, just as the problems of sustainability outlined in chapter 1 pertain to the character of natural systems. The conception of conduciveness to sustainability introduced in chapter 1 offers a more comprehensive basis for cataloging these obstacles than other approaches we have encountered.[2] According to our definition of what is conducive to sustainability, human attributes, practices, norms, settings, structures, cultures, institutions, systems, and policies may all be conducive or not conducive to sustainability, and those that are not conducive to sustainability may qualify as significant obstacles to achieving it. Because human activities are shaped by all of these factors, it should be clear from the outset that progress toward sustainability will require that the offending systems, structures, policies, and so on be directly confronted and reformed through the collective social and political action of citizens. In a world in which 80 percent of global production is controlled by one thousand corporations that wield immense political and social power, an obvious priority would be to "dramatically change the publicly traded, limited liability global corporation, just as previous generations set out to eliminate or control the monarchy."[3] Corporations are inventions of law, and it is an open question how far they could be reconstituted by law to function in ways that are more conducive to sustainability.

Expecting individuals to consume less or buy "green" will not suffice, and coordinated efforts directed toward political action, corporate reform, and collective sustainability projects will only succeed if they are well-directed and informed by sufficient understanding of the nature and variety of obstacles standing in the way of success.[4] The task of this chapter is to provide the basics. Because sustainability is fundamentally a problem of global cooperation, we find it useful to begin by surveying the various obstacles to success in achieving sustainability through individual choices that are not coordinated by government regulation, collective action, or market mechanisms designed to be conducive to sustainability: the three basic forms of coordination. We will then consider obstacles to coordinated collective action to achieve sustainability.[5] There are obstacles common to both individual and coordinated action in this domain, but addressing the former first has the merit of clarifying why coordinated action is necessary as well as some aspects of how it might work. Seeing how it might work and on what basis will reveal the importance of a common understanding of the ethical dimensions of sustainability. If problems of sustainability are largely problems of social coordination that can be solved only through compliance with fair terms or principles of cooperation, then the absence of wide public agreement on such principles is a major obstacle to sustainability.

The importance of coordinated action in the interest of sustainability compels us to note at the outset the limitations of the *demand formula*, according to which throughput or demand is a product of population, per capita consumption, and the footprint intensity of a unit of consumption.[6] This formula is constructed on the premise that the fundamental means to achieve sustainability is to reduce throughput of energy and matter from the environment as resources through consumption of products and services, and back to the environment as waste. The formula offers the options of reducing population, reducing per capita consumption, reducing the footprint intensity of units of consumption, or some combination of these. The distinction between consumption and the footprint intensity of consumption calls attention to the fact that the footprint intensity of units of consumption is systemically determined and only marginally within the control of individuals. This is one reason that uncoordinated individual efforts to promote sustainability are inadequate. The sustainability conduciveness of structural features of societies, including housing, transportation, energy, and food systems, is closely associated with the footprint intensity of units of consumption.

Perhaps even more instructive are the limitations of this formula. It conceptualizes throughput as flows of resources through consumption and back to the environment as waste. This may capture the dominant pattern, but it is somewhat misleading. Some economic throughput takes the form not of consumption expenditure but capital expenditure, and the latter could include investments in natural capital or biocapacity. Other material throughput is used, especially in war, to do the opposite—to destroy natural and manufactured capital. Replanting denuded hillsides, reengineering waterways, and targeting sanitation and energy infrastructures in the conduct of war all have import for sustainability, and none are adequately captured by the ideas of consumption and waste. Further, by treating institutional consumption (by armed forces, businesses, nongovernmental organizations [NGOs], etc.) as an undifferentiated aspect of per capita consumption, the formula invites (though it does not entail) the notion that total consumption is simply a sum of the consumption engaged in by individual natural persons or "consumers."

The well-known IPAT equation—Environmental Impact (I) = Population (P) × Affluence (A) × Technology (T)—is no better in these respects, and worse if it is understood to imply that footprint intensity is purely a function of technology as it is normally understood. Footprint intensity is a function of the entire sequence of acts and processes involved in the generation, delivery, and trailing effects of a unit of consumption, and it follows that improvements in technology as such do not ensure lower footprint intensity. Improvements in petroleum production technology have kept oil flowing as exploration has uncovered fewer significant new oilfields, but at lower EROEI rates, and these lower EROEI rates entail a higher footprint intensity for a gallon of gasoline, unless there are compensating reductions of natural capital consumption in the exploration, production, and delivery processes.[7]

However, if *affluence* and *per capita consumption* are simply terms of art signifying *per capita income* and *technology* is another term of art signifying the *environmental impact of spent income or production*, then the demand formula and IPAT equation are equivalent—and neither accords individual consumer choices and technological improvements the significance they are commonly accorded.

Obstacles to Achieving Sustainability through Uncoordinated Individual Efforts

Substantial progress toward sustainability might nevertheless be achieved through individual efforts to reduce household consumption. A 2009 study found that "direct energy use by households accounts for approximately 38% of overall U.S. CO_2 emissions, or ... approximately 8% of global emissions and [more] than the emissions of any entire country except China. ... [and] 17 types of household action ... can appreciably reduce energy consumption using readily available technology with a low or zero cost or attractive returns on investment, and without appreciable changes in lifestyle."[8] Such actions could begin to reduce emissions immediately, which would be valuable, and they might conceivably occur on a scale sufficient to make a difference in response to informational campaigns, persuasion, and social marketing.

Coordinated efforts to inform and persuade would need to convince the public that problems of climate and sustainability are real and urgent and that the affordable and attractive products marketed as lower carbon or more "green" really are what they claim to be.[9] The incidence of misleading green claims in advertising and marketing is reportedly very high.[10] Deficiencies of public education efforts and enforcement of truth in advertising regulations should thus be numbered among the obstacles to sustainability. As admirable and personally beneficial as individual efforts to reduce consumption may be, collective acts of citizenship are needed to engender an effective public response.

Moreover, adequate reductions of household consumption would require much more than well-informed and motivated individual efforts, and not all consumption is household consumption.

Aspects of Footprint Intensity and Consumption Beyond Individual Control

Units of consumption do not wear their footprint intensity on their sleeves. We typically purchase end products and services with little knowledge of how they are generated and delivered—little knowledge of the sequence of acts and processes the performance of which we have indirectly rewarded. We also may have limited options for disposing of what we no longer want and limited knowledge of what those options entail.

Lacking such knowledge, we are also typically unaware of the full expenditure of natural capital associated with our purchases.[11] Price sensitivity is a prominent aspect of purchases, but price is often a poor signal of the footprint of acts and processes through which a product or service is generated.[12] The fact that a product's or service's full cost in natural capital is typically not reflected in prices is an obstacle to sustainability and one aspect of wider failures of markets to be fully transparent and conducive to sustainability.[13] Energy, water, and food systems offer vivid illustrations of this problem.[14]

Carbon Intensity of Energy Units   An economy built on oil, coal, and natural gas is so obviously carbon-intensive that there can be no disputing the importance of widely adopting low-carbon successor technologies. This is the greater part of reducing footprint intensity, and without the right incentives and support, the research and development needed to create new technologies will not occur. Once new technologies are developed, there are obstacles to widespread adoption or replacement of one energy system with another. Conceptualizing these obstacles as barriers to "commercialization and deployment," a 2007 US Department of Energy report identified the following as the most important: uncertainties about how well the new technology will perform, the "high cost of developing needed expertise," "the absence of established infrastructure," the existence of "support systems" for competing "incumbent" technologies, and—above all—the fact that "GHG emissions from fossil fuel consumption" are "unpriced."[15] Conversion from a fossil fuel economy to a low-carbon successor economy is, in short, very costly in energy, materials, and human capital, fraught with technical and market uncertainties, and not sufficiently profitable without domestic laws—and, sooner or later, international conventions—to limit carbon emissions.

Low-carbon energy technologies are at a price disadvantage that would be diminished or disappear if the full costs of burning fossil fuels were reflected in their price. These costs that are not borne by the buyers and sellers in commercial transactions, called negative economic externalities, are uncompensated harms to third parties. Those third parties are all of us, present and future, who are or are likely to be harmed by the effects of climate disruption, acid rain, or toxic exposures associated with fossil fuels. A *carbon tax* or *cap and trade* (permit) system would aim to capture the negative externalities pertaining to climate in such a way as to make businesses and individuals who voluntarily sell and buy fossil fuels bear those costs of the products themselves.[16] Internalizing the costs in that way would be fair, and it would also be an efficient way to remedy a *market failure* (meaning an inefficient allocation of costs and benefits by a market) that presents a barrier to the adoption of less damaging technologies—technologies that would lower the footprint intensity of units of energy consumed.[17]

Internalizing the full environmental costs of fossil fuels in their prices would have widespread effects on the prices of everything for which manufacture, distribution, or operation currently depends on those fuels, creating incentives that could be remarkably efficient in yielding carbon-lowering innovations. Structural obstacles to lower-carbon living might in that way be overcome, and the carbon-sensitive price signals would take the guesswork out of carbon footprint reduction for everyone. The effect would be not only reduction of the footprint intensity of a unit of energy consumed, but also a reduction in the energy units devoted to other consumption. The footprint intensity of those other things consumed matters too, of course.

Footprint Intensity of a Daily Commute   Is a daily commute from home to work and back a unit of consumption? Let's suppose it is. The footprint intensity of a daily commute is largely a function of the footprint intensity of energy and structural aspects of urban settings and their transportation systems. What is the distance from affordable and desirable housing to workplaces? How extensive is public transportation, and is it used by enough passengers to yield a substantially lower footprint per commute than that entailed by use of personal vehicles? To what extent is walking or bicycling feasible? Are there sidewalks for pedestrians? Safe bicycle lanes? Options for combining bicycle and tram? Can shops be reached on foot, or are they concentrated on the suburban fringe? Many choices regarding urban design and thereby sustainability are no less public choices than the imposition of a carbon tax would be. Therefore, although it is useful to encourage individual lifestyle changes conducive to sustainability, it is also important to understand the role of systems, settings, and structures in shaping individual choices and to encourage cooperation in public action and public-private partnerships to make these determinants of individual choice more conducive to sustainability.

Footprint Intensity of Food   What are the full costs in natural capital of a unit of food, taking into account not only the carbon intensity of fertilizer, livestock GHG emissions, and fuel for machinery, but also the biocapacity loss entailed by deforestation, soil loss, fertilizer and livestock waste runoff, and dispersion of herbicides and pesticides? Such costs are further aspects of the footprint intensity of a unit of food and further externalities imposed on third parties whose opportunities are diminished by loss of biocapacity. The merits of internalizing these costs in price are no less evident than in the pricing of GHG emissions, but internalizing them would require not just a carbon tax but a *biocapacity tax* (assessed in order to limit biocapacity loss or promote ecological sustainability). If the goal is throughput sustainability, then limits on fossil water extraction would also be required. Would such steps serve to lower the footprint intensity of units of food and in that way be conducive to sustainability? Yes, though they would probably do so largely by inducing dietary substitutions. Sustainability will likely require *different* consumption—changing patterns of consumption, not just less consumption.

Institutional Consumption   Institutional consumption by businesses and governments is significant, and the extent of its limitation by market mechanisms, accountability to corporate boards and shareholders, and mechanisms of political accountability is unclear. The regulation of business consumption would presumably depend to some extent on characteristics of the market in question, including how small the field of players is, and the public regulatory environment that influences market conditions. The latter is directly political or a matter of public effort exerted through acts of government. Limitations of consumption by public institutions would presumably be most conducive to sustainability in democracies that are guided by sufficient understanding of sustainability, able to invest adequately in well-conceived efforts to promote sustainability, and well-ordered (i.e., transparently respectful of principles of justice and evidence).[18] A notable aspect of corporate institutional consumption, to which we shall return, is how much of it is devoted to frustrating public understanding and oversight of corporate activities. This is consumption not only beyond the control of individuals, but devoted to frustrating public efforts to ensure that corporate activities are compatible with the public interest.

Cultural and Social Structural Obstacles to Consuming Less

Invention Is the Mother of Necessity[19]   Some consumption is dictated by avoidance of deprivation of what have become necessities within one's society, as innovations have made formerly rare forms of consumption widely available. *Necessities* in this sense include, as Adam Smith famously

wrote, "what ever the customs of the country renders it indecent for credit-
able people, even the lowest order to be without. ... [such that they] would
be ashamed to appear in public without them."[20] We referred to such consumption in chapter 1, in connection with norms of respectability and
cause for shame or loss of face. The contextual significance of such consumption for social standing and self-respect makes it essential to living
well in the relevant social context, but it may be in other respects a luxury
that could be given up without significant sacrifice of well-being. Such consumption is often subject to a dynamic of inflation: "It would seem that
wherever there is material or technological progress, new goods will gradually assume the role of 'signifying decency' ... This is important, for it suggests that technological progress combined with the need for self-respect
tend to up the consumption ante."[21]

Escape from poverty is essential to the satisfying fulfillment of human
possibilities, but beyond a threshold that affluent countries of the North
have long since passed, rising levels of affluence offer mostly an illusion
of progress. Although *relative* social position has substantial impact on
well-being, rising affluence may do little more than shift thresholds of
respectability and success further out of reach for many.[22]

To the extent that individuals associate loss of face with failure to satisfy
social expectations to consume, and those expectations are not independently warranted by the contributions of such consumption to living well,
such expectations (or norms of consumption) are obstacles to uncoordinated individual efforts to reduce consumption.

Culture   Cultures evolve with invention, but they may also preserve bonds
of attachment to practices and norms of consumption, even as the forms of
consumption involved become far more destructive. Jared Diamond depicts
the Greenland Norse as having perished largely due to their cultural attachment to livestock in an unsuitably fragile land and their resistance to eating
fish that were abundant and close at hand (as evidenced by the almost complete absence of fish in Norse archeological sites).[23] Their story is a telling
illustration of the risks people will endure to preserve their culture and a
culturally defined identity.

The practice of exchanging gold in weddings makes for an interesting
comparison: It remains a symbolically important ritual in which countless
people participate, with little or no comprehension that the footprint
intensity of gold is far beyond what it was in years past. The world's
remaining gold ore is of such marginal quality that the extraction of
a single ounce—the gold in one ring—requires thirty tons of ore and

commonly involves the use of a cyanide leaching process that contaminates the mountains of debris left behind.[24] At the more environmentally conscientious open pit mine at Batu Hijau in Indonesia, 250 tons of rock and ore are required to produce an ounce of gold. Toxic chemical tailings are piped out to sea, and the discarded rock is spread across formerly pristine rain forest. The mine alone is large enough to be visible from space.[25] With little more than microscopic specks of gold left in the ground, the engineering feats required to keep the supply flowing are stupendous and commensurately destructive. The use of gold is a noteworthy example of a cultural obstacle to sustainability, and our circumstances invite the examination of such obstacles in the spirit of preserving what is best in the cultures that are important to our identities and well-being. Do we need to calculate beauty return on beauty invested (BROBI) ratios to be cognizant of the acts we reward in buying gold?

Strategic Consumption and Positional Advantage   Individuals and enterprises tend to seek their own good by striving for advantage in competing for desirable positions and striving to enhance the advantages of the positions they have. They engage in *strategic consumption* pursuant to *positional advantage*, and such consumption is subject to inflation and may be otherwise unnecessary. It may be more avoidable than the contextually necessary consumption "signifying decency," but in a system in which a great deal is at stake in not falling from a higher social stratum to a lower one, it may be all but irresistible.

For an illustration of the problem, consider the complete deforestation of Easter Island, which resulted in large measure from the status competition between the chiefs of its rival clans. By competing to erect the tallest stone heads (*moai*) representing high-ranking ancestors, these chiefs oversaw the clearing of trees for the manufacture of sleds, ropes, levers, and "canoe ladders" used to transport the carved stones from a central quarry and erect them on platforms along the coast. The tallest stood thirty-two feet and weighed seventy-five tons, but nearly half of the 887 *moai* inventoried by researchers remain in the quarry in various stages of completion, suggesting an industry that was growing rapidly and collapsed. Twenty-one species of trees were destroyed, leaving a starving and collapsing population without fuel, native bird species, the materials needed to construct canoes for deep-sea fishing, or trees to regulate the absorption of rain and control soil erosion.[26]

To the extent that unsustainability is fueled by a high-stakes pursuit of competitive advantage, it is a problem of collective choice, the solution to which would require norms and policies that limit the intensity and destructiveness of competition.

Advertising

Advertising exists to induce people to purchase goods and services they would otherwise not purchase. It is possible to imagine a world in which advertising results in fully informed and rational behavior, characterized by regard for the true benefits and full costs of the goods and services, but we live very far from that world. We would do well to acknowledge that advertising is an obstacle to sustainability. It induces people to purchase goods and services they do not need, and it contributes to a culture of consumption that makes people less happy and more burdened with anxiety and debt.

Advertisers have learned from the psychologists they hire that anxiety makes people less happy and induces acquisitiveness, so they formulate messages designed to induce anxiety, knowing the messages will both stimulate sales and make people less happy. Research on well-being has shown that the more advertisements people view, the less happy they are. The more value they attach to wealth, the less happy they are. The more they shop and buy, the less happy they are.[27] Advertisers aggressively target children, bombarding them with messages known to be bad for them; advertising messages and the materialistic values conveyed produce not just anxiety, but depression, negative self-image, a higher incidence of risk-taking, and conflict with parents.[28]

Whatever is being sold, a broad message of advertising is that participation in markets as a consumer solves one's problems and changes one's life for the better—a message continually reinforced by wider industry praise for markets and undermining of collective and government action, mediated by front organizations, think tanks devoted to sophisticated endorsement of industry agendas, and sponsorship of political spokespersons, all in the interest of enfeebling public oversight of commercial activity, weakening the bargaining position of labor, and commercializing what is not yet commercialized. The task of conceiving of ourselves as citizens sharing public interests that can only be secured collectively is made that much harder but all the more essential.

Reproductive Freedom

"There is a lot of fear about—even outright hostility to—talking about population and environment in the same breath," Laurie Mazur and Shira Saperstein write in their afterword to a major work on population, environment, and justice.[29] The fear is understandable, but the importance of acknowledging population growth as an obstacle to sustainability is undeniable. Many things of importance to sustainability become harder with population growth. The task of developing the human capital essential to managing socioecological systems is easily overwhelmed by population growth, and so too is the alleviation of poverty that is an essential prerequisite to individuals taking a long-term view of resource systems. Population growth in the global North and wealthy enclaves of the South plays a far greater role in the growing human footprint than general population growth in the South does, of course, but poverty and insecurity are obstacles to a long-term perspective everywhere. From an ethical standpoint, it is the vulnerability of the poor and the weakness of reproductive rights for women in much of the world that are of primary concern. From a footprint-reduction standpoint, the view widely held in international development and human rights circles is that poverty reduction, universal primary education for girls, and universal reproductive rights for women offer the most promising path to stabilizing population.

The 1994 Cairo Program of Action, endorsed by 179 states at the International Conference on Population and Development in Cairo, defined reproductive rights in an international policy document for the first time: "Reproductive rights ... rest on the recognition of the basic rights of all couples and individuals to decide freely and responsibly the number, spacing and timing of their children and to have the information and means to do so, and the right to attain the highest standard of sexual and reproductive health. It also includes their right to make decisions concerning reproduction free of discrimination, coercion and violence, as expressed in human rights documents."[30] The legitimacy and effectiveness of any policy initiatives designed to stabilize population would almost certainly depend on consistency in respecting these and related human rights. Women generally choose to delay childbearing and have fewer babies when they are able to obtain education and have the opportunity to make such choices.[31]

Lack of knowledge and reproductive self-determination are arguably the salient obstacles to sustainability where population stability is concerned. There is strong evidence that universal elementary education for girls is an excellent investment in enabling women to make responsible reproductive decisions for themselves, and there has been good progress in closing the global gender gap in elementary education and ensuring universal access to it. Enrollment of girls in elementary education was only two-thirds of what it was for boys in 1990 but had risen to match that of boys by 2010; there are now more children enrolled in elementary education than there are primary-school-age children in the world.[32] This is encouraging, though other barriers to reproductive freedom may still constitute an obstacle to sustainability.

Systemic Failures of Public Knowledge

A further obstacle to sustainability being achieved through uncoordinated individual efforts is the deficiencies of systems of *public knowledge*, systems by which inquiry is advanced in accordance with the most objective and reliable methods of investigation available, and knowledge obtained through such methods is broadly disseminated in the public interest.[33] We are all dependent on others for most of what we know, and such *epistemic dependence* is a basic aspect of the human condition. It subjects us to considerable risk when our trust is misplaced, but our ability to pool our knowledge offers immense advantages when we are efficiently organized to distinguish truth from error and share important truths. Open societies pursue those advantages by investing in scientific inquiry and institutions through which such inquiry's results are communicated to policy makers and the public. An essential aspect of this enterprise is establishing a *division of epistemic labor* in which diverse forms of epistemic expertise and veracity are cultivated and the resulting intellectual virtues are reliably harnessed in the public interest. This requires a system of institutions that assign recognizable epistemic authority on the basis of epistemic merit, facilitating public investment of trust in those who are most epistemically reliable. It also requires that the conditions of work in these institutions align with the fulfillment of epistemic merit in *good work*—in discernment and communication of important truths.[34] Entities such as national academies of science and the Intergovernmental Panel on Climate Change have been designed to gather research that has survived peer review, identify points of consensus relevant to the public interest, and communicate what they have identified as known and important. These institutions serve the public interest when they enhance the public voices of experts chosen on the basis of epistemic merit, but being heard and heeded is another matter.

Scientists have overwhelmingly understood their role in the scheme of public knowledge to be limited to doing good research within their specializations, publishing it in peer-reviewed journals written for other scientists, and training future scientists. Although many scientists are also educators, they are rarely perceived as having a responsibility to communicate their findings directly to the public and policy makers in terms suitable to guiding important decisions, nor have they often sought out interdisciplinary collaborations in order to serve the public interest. Sustainability evidently requires better public communication of science and a greater attunement of research agendas to problems that cut across existing specializations, so these aspects of the prevailing norms of professional responsibility present obstacles to sustainability. How will the project of public knowledge respond to the problems of a global public and fruitfully inform public deliberations in the present age without "a new kind of [global-change] science ... better able to juggle empirical, technical, political, ethical and other matters at the same time"?[35] Scientists are debating this question among themselves.[36]

Educational Institutions An obvious function of educational institutions at all levels is to disseminate knowledge that is vital to the public interest, and this plainly should include what is known about sustainability and how it might be achieved. The fact that dissemination of such knowledge scarcely occurs in most schools in the United States is another obvious obstacle to sustainability. Although sustainability is now mentioned in national standards for science education, the conditions for an education in sustainability actually occurring are generally lacking. Curricular integration is rare, and high-stakes testing encourages a focus on basic skills and recall of facts at the expense of interdisciplinary understanding, critical thinking, and problem solving of the kinds that would be essential to sustainability education. American educators are not well-prepared to engage in significant teaching related to sustainability, and there is a dearth of resources through which this lack of preparation can be efficiently overcome. Barriers to the emergence of adequate sustainability curricula in the arena of higher education begin with the withering of general education in favor of fragmented, specialized, and career-focused programs of study, and they include norms of academic life that have allowed college faculties to too often ignore their collective responsibility to provide students with a coherent general education.

Mass Media Wider dissemination of important truths known to communities of experts is unfortunately hampered by the limitations of popular media. The idea of "balanced" journalism has too often resulted in reporting that undercuts scientific consensus by pairing it with critics who are not scientists, not experts in the relevant science, or have not done research of the relevant kind in many years and are no longer active, respected members of the relevant scientific community. Consensus science is occasionally overthrown (i.e., replaced by a new consensus), but consensus within the relevant body of experts remains the public's best guide to the truth, and the existence of a few outliers even within the relevant community of experts does not justify the common journalistic practice of giving the "skeptics" equal time. Skeptics within the relevant community of scientists will have had their chance to influence the consensus, and the idea that they deserve a second chance with the lay public or that the public is well-served by the presentation of a "controversy" it cannot competently judge is misguided. Skeptics who do not belong to the relevant scientific community and who are truly motivated by the public interest should test their ideas in the forum of that community, trusting it to be no less dedicated than they are to advancing knowledge in the public interest.

Journalists do not need to be experts on the science they report to get it right, but they do need to understand the nature of science, the role of epistemically reliable institutions in the practice and communication of science, and the difference between (1) "he said, she said" disputes engaged in by comparably positioned witnesses to an event and (2) critiques of scientific knowledge by people who make careers of delaying government action on matters of public health and environmental protection.[37] Failures of science reporting are systemic in US media and are exacerbated by the segmentation of media markets by ideological orientation and greater public trust in ideologically compatible news—factors that contribute to a persistent gap between what climatologists know and what the public knows.[38]

Corporate Public Relations   Public relations revenues in the United States have grown rapidly in recent years, and much of that growth has been devoted to "greening" corporate images.[39] These efforts contribute to the public being badly misinformed about the public health effects of pollution, the strength of scientific consensus on climate change, and what defines good science. The strategies used are diverse, well-funded, and sophisticated, and they often replicate those of the tobacco industry. The fossil fuel industry in particular has devoted vast resources to undermining public understanding of the scientific consensus on climate change, a consensus that industry scientists have privately accepted for at least two decades.[40]

For many years, the tobacco industry waged a secret campaign to dispute the evidence that tobacco is addictive and causes cancer and other diseases—evidence accepted as conclusive by their own scientists—waging that campaign through public relations firms, front organizations such as The Advancement of Sound Science Coalition (TASSC), and industry scientists posing as independent guardians of "sound science" in the public interest. The aim was to create an illusion of scientific controversy through distortions of standards of evidence, misleading presentation of industry-sponsored research, and suppression of competing research through restrictive business-university partnerships, legal maneuvering, and attacks on independent scientists. The strategies and language of so-called sound science—invented by a public relations firm hired by tobacco industry lawyers—evolved into a broader campaign against unwelcome scientific inquiry, waged substantially through industry sponsorship of antienvironmental think tanks, climate change–denier websites, and sham climate conferences at which scientifically debunked arguments are endlessly recycled.[41] James Powell, the executive director of the National Physical Science Consortium, wrote in 2011 in response to strengthening scientific consensus on climate change: "In 1997, Amoco, British Petroleum, DuPont, Ford, and Shell resigned from the Global Climate Coalition [a climate denial organization that operated from the offices of the National Association of Manufacturers] to join ... [the environmentally responsible] Business Environmental Leadership Council," but between 1998 and 2005, ExxonMobil contributed "$16 million to more than forty organizations that deny global warming."[42] As documented in the Union of Concerned Scientists 2007 exposé, *Smoke, Mirrors, and Hot Air: How ExxonMobil Uses Big Tobacco's Tactics to Manufacture Uncertainty on Climate Science*, ExxonMobil has relied on the same strategies and some of the same key players as the tobacco industry, as it has "manufactured uncertainty, laundered information, promoted pseudoscience, called for 'sound science' to divert attention from incontrovertible scientific evidence, and used its access to government to deny and delay action."[43] Yet it is now clear that ExxonMobil has not been alone in funding obfuscation. Although Shell and other companies began acknowledging in 1998 that fossil fuel emissions are a cause of climate change, "fossil fuel companies' broader campaign to sow confusion continued" through the American Petroleum Institute (an industry trade group), Western States Petroleum Association, and other entities.[44]

The distinguished philosopher of science Philip Kitcher is unflinching in his assessment of the difficulty we face: "Much of our thinking about the value of free discussion presupposes an individualistic epistemology that takes citizens to suffer from directly remediable ignorance—their natural wit enables them to see Truth as the winner in the encounter. In the contemporary world, that presupposition is so far from reality that it is hard to credit it even as an idealization. ... When the channels through which information is distributed to the voting public systematically distort findings that have achieved full international consensus ... irremediable ignorance is bound to flourish."[45] Climate "skeptics" exploit the presumption that common sense is an adequate guide to truth, with such sophistries as "nature is, as any reasonable person might suppose, dominated by stabilizing negative feedbacks rather than destabilizing positive feedbacks."[46] This is a good illustration of the fact that in matters as complex and contested as climate change, common sense is not enough. Well-functioning institutions of public knowledge are essential.

Psychological Traits and Mediating Social Norms and Practices

There are common cognitive, motivational, and affective traits that may be obstacles to achieving sustainability: cognitive dispositions that interfere with rational uptake of evidence, difficulties in understanding complex systems, aspects of motivation that contribute to excessive consumption and procrastination, and emotional responses that contribute to denial. Such traits might be overcome to some extent by well-designed educational interventions and institutional arrangements that beneficially structure choice sets and social responses to troubling emotions. To the extent that such interventions are needed, the traits in question are further obstacles to achieving sustainability through uncoordinated personal efforts.

Psychologist Robert Gifford has attempted to catalog the psychological obstacles to reductions of household consumption by sorting them into seven categories:[47]

1. *Cognitive limitations:* distraction by competing concerns, lack of knowledge, irrational response to uncertainty, discounting of evidence, optimism bias, perceiving oneself as unable to influence events
2. *Ideologies:* conflicting worldviews, faith in superhuman or technological saviors, status quo bias
3. *Influence of social norms:* for example, norms of consumption that operate through cues, emulation of respected others, and the role of social comparisons in perceived entitlement
4. *Sunk costs:* in the form of financial investments, established habits, and vested interests
5. *Discredence:* mistrust, denial, and oppositional reaction to unwelcome messages
6. *Perceived risk:* irrational risk aversion in contemplating novelty
7. *Tendencies to limit changes of behavior:* limiting to those that are the least costly

One would not need to follow debate about climate change in the United States for long to encounter at least a few of these barriers, and it would not be unusual to find several of them in combination: ignorance of the extent of consensus about climate disruption among bona fide climatologists; ideological distrust of environmentalists, the United Nations, and defenders of government regulation and taxes; vested interest in preserving the business-as-usual status quo; spatial discounting, or the tendency to assume the problems of climate disruption will occur primarily elsewhere; and misplaced faith in the capacity of markets to generate replacements for anything our way of life may destroy. Denial rooted in fear, guilt, and a need to feel in control of one's fate, and the human tendency to give excessive weight to evidence that supports one's beliefs and insufficient weight to evidence that undermines one's beliefs (known as *confirmation bias*), may both be factors in the difficulty many people seem to have in imagining that the circumstances of our lives might be very different before long. The same traits may contribute to the difficulty many have in recognizing that it is a form of *magical thinking* (or ideologically based *faith in technological saviors*) to assume markets will reliably identify and scale up technological forgiveness of our ecological debts, always in time to assure smooth sailing.

Mediation of Denial by Social Norms and Practices Denial is not simply an individual psychological, self-protective response to troubling feelings of helplessness and guilt, but a collective *social process*, mediated by cultural norms and practices that regulate *attention* (hence distraction by competing concerns), *emotion*, and the boundaries of acceptable *conversation*.[48] As such, it may play a powerful role in mediating personal response to problems of climate and sustainability, even among those who understand and are concerned about these things.[49] "We learn what to see and think about from the people around us," Kari Marie Norgaard writes in a revealing study of public silence and inaction in the face of wide understanding and concern about climate change in Norway.[50] She observes, much as we have in our own teaching, that learning about climate change and other aspects of unsustainability is *disturbing*. The anxiety and feelings of vulnerability, guilt, and helplessness it provokes are not only hard to live with, but destructive of personal health and well-being if they fester and are not relieved through constructive action to address the problems.[51]

Whether such corrosive feelings are relieved through constructive engagement or through denial is evidently regulated in large measure by the practices and norms of communities. The scarcity of communities that invite and facilitate constructive engagement in action conducive to sustainability may be both the most important of all the obstacles to sustainability and the easiest to overcome by individuals willing to take some initiative. The most effective forms of communication about climate change and related matters "engage people in groups, and give them opportunities to develop their understanding and their narratives about consumption in dialogue together,"[52] and the social mediation of denial may be one important reason why. Enabling people to overcome their troubling emotional responses to evidence of climate change and unsustainability requires social settings that provide opportunities for both dialogue and meaningful action to promote sustainability.

Cognitive Traits What Gifford has assembled is not exactly a catalog of psychological barriers, but instead a mixed bag of inherent psychological traits and externally shaped states of mind that play roles in failures to understand climate disruption and contribute sufficiently to its mitigation. Confirmation bias is clearly a psychological trait and barrier to rational belief, but lacking knowledge or possessing false antecedent beliefs that are reinforced by confirmation bias is a mental condition largely attributable to the social, cultural, and institutional contexts in which individuals acquire ideologies, worldviews, and knowledge. If the psychological dispositions were distinguished from the externally shaped states of mind, the former could be characterized as obstacles to sustainability insofar as they are defects of rationality, reasonableness, or both—*rationality* in judging evidence about climate and sustainability and deciding what to do to protect one's own interests and *reasonableness* in doing one's fair share to promote sustainability.[53] General limitations of rationality in responding to evidence should also be distinguished from the difficulties people generally have in understanding complex systems involving stocks and flows, time delays, and feedback—features that are characteristic of climate regulation and ecological systems generally.[54]

These general cognitive deficiencies of response to evidence and specific difficulties in understanding complex systems invite educational response in the form of instruction in critical thinking (to enable students to monitor and improve the quality of their own thinking) coupled with understanding of environmental systems and sustainability-related aspects of human systems, including the ways in which cultural norms and practices limit appropriate response to problems of sustainability.

Motivational Traits   Gifford's category of *influence of social norms* is similarly a mixed bag of norms and propensities to emulate and compare oneself with those in favored social positions. Other motivational traits addressed in the sustainability and environmental literature include acquisitiveness and anomalies of preference that lead to procrastination or a failure of timely action to secure a preferred outcome.

Acquisitiveness may vary as a function of insecurity and other contextual factors, as psychologists have argued, yet also be a dispositional trait that was adaptive for human survival and reproductive success, as biologists have argued.[55] Neither perspective implies that acquisitiveness is always an obstacle to sustainability, but in present circumstances (a pervasive culture of consumption, the impact of widening inequality on feelings of satisfaction and insecurity, and so on) it would seem evident that certain aspects of human acquisitiveness are obstacles to sustainability.

Discounting the value of future rewards, or giving more weight to current rewards than future rewards, is a widely observed human tendency, and it gives rise to reversals of the rankings of preferences.[56] These reversals, or transitory spikes in the strength of desires or preferences, occur when pleasant but less preferred objects of desire are close at hand or temporally proximate—such as when the presence of ice cream elevates its desirability over avoidance of diabetes. Because the spikes are transitory, the satisfaction of the desires is often tainted by regret. Once the initial euphoria has worn off, coming into possession of desired objects often leaves people less happy than they were beforehand.[57] More troublesome is the pattern that emerges when a series of such preference reversals incrementally preclude achievement of a more highly preferred goal. A smoker may have a strong preference to not be a smoker at all, hence not have the next—or any other—cigarette, yet find the desire for each next cigarette to be her strongest desire at each choice point. Similarly, for shopping addicts whose dominant preference is to escape debt and people who see the wisdom of saving for retirement or contributing to sustainability but delay getting started, at each choice point they may prefer current consumption to a specific investment in the future. From many such momentarily rational choices, a monumental irrationality may result: the irrationality of failing to avert a disaster one prefers to avert by allowing oneself to be mastered by desires one would have preferred not to indulge.

Understanding that some irrationality is structural, we can commit ourselves, individually and collectively as citizens, to arrangements that alter the structure of our moment-to-moment choices: automatic savings plans for retirement, a national pension system, a universal carbon tax, and the like. Such arrangements may be thought of as strategies of precommitment that reshape the architecture of choice in beneficial ways.[58]

Obstacles to Coordination

There is widespread agreement that protection of the global atmosphere, a global *commons* or *common asset*, requires a system of global governance—a body of international law, essentially—and that anything less comprehensive will fail, because global cooperation requires assurance that individuals in every region are doing their fair share. A *tragedy of the commons* is said to occur when there is open access to a resource, free use of that resource by multiple users damages or uses up the resource, and—perceiving this—users find it rational to continue or accelerate their use so as to not lose out to competing users, even though cooperation would provide them a better outcome in the long run.[59] The *theory of collective action* suggests that external enforcement of rules or *coordinative principles* is required to solve such collective action problems, because without such enforcement individually rational behavior will be a decisive obstacle to mutually advantageous cooperation. When the number of individuals involved in selecting coordinative principles is so large that face-to-face negotiations cannot include everyone, there must be leaders, and once these ground rules of cooperation are determined there must be neutral adjudication of conflicts and common enforcement mechanisms. The theory of collective action and social contracts thus identify an absence of competent government authority—legislative, judicial, and penal authority—as an obstacle that must be overcome to achieve sustainability. How this authority would be developed and structured—how centralized or how distributed it would be—remains unclear.

Efforts to negotiate an international treaty to limit GHG emissions have been predicated on this understanding of the situation, but those efforts have encountered other obstacles. One such obstacle is the strategy of brinksmanship, whereby countries delay cooperation in the interest of obtaining the best possible terms for themselves. This may delay cooperation indefinitely if there is not a mutually acknowledged deadline for coming to agreement. Asymmetries of historical responsibility and vulnerability between wealthy countries of the global North and poorer countries of the global South are also likely to complicate negotiations. Then, once there is an agreement, an international treaty only takes effect when a specified majority of signatory nations ratify and give it effect through enabling legislation within their own jurisdictions. Political leaders must face their domestic constituencies in defending passage of the laws, and without favorable political conditions at home they may be reluctant to put their own careers on the line. The fact that enforcement of international environmental law is through domestic laws is also one of several obstacles to the global trust that is essential to reaching agreement on a set of global coordination principles. It may be easy to doubt how effectively other governments would monitor and enforce compliance. If all of these obstacles were overcome, obstacles to success in administering the agreements still would remain: There would need to be effective monitoring and enforcement sufficient to ensure a high level of compliance, which is all but impossible without widespread acceptance of the legitimacy or fairness of the terms of cooperation, and rules would need to be adapted to local and evolving circumstances. Having duly noted these specifics, one might reasonably conclude that the greatest obstacle to success is humanity's lack of experience with meeting a challenge as big—in scale, peril, and complexity—as human-induced climate change. The benefit of prior experience would be in having learned lessons that could now be applied.

These are the obstacles to comprehensive and effective global treaties to secure the structural conditions for sustainability, and it is through a *multiscale approach* that they are most likely to be overcome. Elinor Ostrom, a Nobel laureate in economics and pioneer in the study of self-organizing management of common resources, has identified the factors favorable to local and regional self-organization and argued forcefully that there are advantages in organizing cooperatively at a variety of scales to address climate change and other environmental collective action problems. The research she and others have conducted over several decades has revealed considerably less empirical support for the theory of collective action than anticipated, while demonstrating that "a surprisingly large number of individuals facing collective action problems do cooperate. Contrary to the conventional theory, many groups in the field have self-organized to develop solutions to common-pool resource problems at a small to medium scale."[60] Sufficient assurance that others will cooperate evidently is present in diverse settings in which there is no centralized enforcement of rules so that individuals find it rational to cooperate by voluntarily complying with norms and participating in the enforcement of those norms through social networks. In light of this research and the failure of centralized governance that had replaced successful customary management of common resource systems, the developing world in the past decade has experienced a broad shift to decentralized environmental governance through local common property arrangements.[61]

Ostrom identifies the following factors as among the most important that increase the likelihood of successful self-organization to manage common resources: individuals share an accurate understanding of the function of ecological systems and the costs and benefits of their own actions; they regard the system or resource as important and take a long-term view of its value in their decisions; it is important to each of them to have a reputation for being a "trustworthy reciprocator"; they can communicate with each other; they are able to engage in "informal monitoring and sanctioning" and consider it appropriate to do so; and "social capital and leadership exist" and are associated with "previous successes in solving joint problems."[62] Two further factors of interest are that users who "share moral and ethical standards regarding how to behave in groups they form ... and have sufficient trust in one another to keep agreements" are more likely to cooperate, and cooperation is more likely when users "have full autonomy at the collective-choice level to craft and enforce some of their own rules."[63] The absence of these and other features that make cooperation more likely may be considered obstacles to cooperation. Accurate understanding, communication, shared norms, trust, autonomy, and leadership are notable, because they are all potentially responsive to education of one kind or another.

Having noted the feasibility of self-organization at local and regional scales, Ostrom makes three important arguments for proceeding with a multiscale approach, acknowledging that global agreements will be indispensable. The first is the relative *speed* of organization at these levels, compared with the global scale. The second is the *efficacy* of distributed administration. The third is the importance of multiscale self-management to the *legitimacy* of any global regime of environmental governance that emerges. This third argument is very important in our view; government policies and laws are not self-implementing and cannot be imposed on unwilling populations through force. Voluntary compliance based on understanding and perception of policies as fair is essential, because without it the costs of enforcement will be both unmanageable and unjustifiable.[64]

Participation in *polycentric and multiscale self-organization* (i.e., organizational activity that is centered in many locations and operates on multiple geographic scales) can and does occur.[65] It often begins and is most effective locally, and much of it is being mediated by social networks, nongovernmental organizations (NGOs), and social movement organizations (SMOs).[66] It may be focused on direct resource management, community and regional sustainability projects, negotiation and political action to change corporate behavior, political action to influence the policies of one's own government, or political action to influence the negotiation of international treaties.[67] Such participation is proving foundational to the formation of a global civil society that can hold international bodies accountable, provide representation for diverse global stakeholders, and engage in transnational public discourse bearing on deliberations pursuant to environmental governance. Theories of the democratic legitimacy of global environmental governance assign varying weight to accountability, representation, and participation in transnational spheres of public discourse, but such theories seem to concur in regarding these as the factors important to legitimacy.[68] What is important to legitimacy from our point of view is that wide participation in polycentric and multiscale self-organization sufficient to create a functional global civil society is an essential and efficacious precondition for the possibility of a global system of environmental governance that would be understood and perceived as fair.

The argument from the efficacy of distributed administration rests on this and several other well-grounded claims: that large units of government often lack the resources to effectively manage resource problems, reliance on local monitors is essential, adaptive tailoring of management to local circumstances is important, trust is more easily established in smaller scale systems, local cooperation can be built on the local benefits it generates in energy savings and other benefits, and a multiscale approach permits experimentation that can inform subsequent efforts—making up incrementally for humanity's lack of experience in solving problems on the scale of global climate change. Ostrom writes:

The literature on global climate change has largely ignored the small but positive steps that many public and private actors are taking to reduce greenhouse gas emissions. A global policy is frequently posited as the only strategy needed. ... Positive actions are underway at multiple, smaller scales to start the process of climate change mitigation. ... Building a global regime is a necessity, but encouraging the emergence of a polycentric system starts the process of reducing greenhouse gas emissions and acts as a spur to international regimes to do their part.[69]

Writing with the benefit of many years of involvement in national and global environmental governance, James Speth and Peter Haas have made the related argument that concrete steps toward footprint reduction by households and citizens acting collectively through NGOs play important roles in advancing sustainability. By demonstrating a willingness to voluntarily bear some of the external costs of their own consumption, informed citizens create markets for more environmentally responsible products and

practices and they demonstrate the political feasibility of stronger government action. An encouraging number of corporate leaders have responded
to market expectations and opportunities with innovations and voluntary
environmental standards and labeling. The initial reductions in the use of
ozone-depleting chlorofluorocarbons (CFCs) came about in this way, as did
forestry and fishery environmental sustainability certification standards.
Political leadership on climate and sustainability would undoubtedly
respond to higher public expectations and stronger pressure for action,
combined with multiscale voluntary and government-mediated efforts that
demonstrate a desire for action.[70] Changes in individual behavior and
nongovernmental self-organization are not enough in themselves to solve
problems of sustainability, but they can make crucial contributions and
offer a vital bridge to the government action and strong global treaties that
will be needed by building social capital, shifting norms, and demonstrating a public readiness to adapt in the interest of sustainability.[71]

Meanwhile, after two decades of failures to compel GHG emissions
reductions, the system of cooperation that may be emerging from the 2015
Paris agreement on climate change looks less like a global social contract
that will be enforced by an overarching regulatory body and more like the
systems of self-organized environmental management that Ostrom studied.
If countries are able to engage in distributed monitoring and sanctioning,
this may be the most promising way forward. However, Ostrom's observations make it clear that although we must build and continue to rely on
forms of polycentric and multiscale self-governance, the adoption of an
international regime of enforceable targets based on fair terms of cooperation will be crucial, and these targets must govern each of the key planetary
boundaries, not just climate stability. One way in which this might work
would be to condition membership in a "Planetary Boundaries Club" of
countries on adherence to agreed-upon emissions reductions and other
measures and to penalize nonmembers through "uniform percentage tariffs
on the imports of nonparticipants into the club region."[72]

Conclusion

We have surveyed a wide range of obstacles to sustainability in this
chapter and have done so in order to understand what kinds of measures
might be efficacious in pursuing sustainability so as to preserve future
opportunity. Our basic conclusion is that problems of sustainability are
largely problems of social coordination that can only be solved through
compliance with fair principles of cooperation, whether such cooperation

occurs through nongovernmental collective action or through acts of government. Beyond this, we have accepted the view that both progress toward sustainability and the legitimacy and efficacy of the policies, treaties, and legislation that will be required will depend on widely distributed participation in polycentric and multiscale self-organization. Finally, we have foreshadowed some of what we will say in chapters 4 and 6 about just and sustainability-conducive institutions and the ways that better education can contribute to sustainability. We have noted the important role for government policy in bringing price incentives in line with sustainability, but our focus is the normative principles that should guide policy, not the mechanisms of policy.

The task of chapter 3 is to identify the basic ethical norms and fair principles of cooperation that should guide collective and government efforts to create and preserve opportunities to live well. We wrote in chapter 1 that no climate treaty will be adopted that is not perceived as reasonably fair to signatory countries, nor will such a treaty be implemented through domestic enabling legislation and widely obeyed unless it is widely perceived as fair within signatory nations. The same is true of treaties governing all the other planetary boundaries that will matter to future well-being. Theories of justice offer vital perspectives on these matters. In the next chapter, we will provide a brief introduction to developments in this sphere and outline an approach that reflects research in the psychology of well-being and motivation and supports long-term comparisons of opportunity to live well.

3   Sustainability Ethics and Justice

We argued in the previous chapter that problems of sustainability are largely problems of social coordination that can be solved only through compliance with fair terms or principles of cooperation. The absence of wide public agreement on such principles is a major obstacle to sustainability, and the task of this chapter is to offer starting points for overcoming it. We identify the basic elements of an ethic of sustainability, relying on the conception of sustainability presented in chapter 1. With the basic elements of this ethic in hand, we will then focus specifically on the responsibilities of governments, provide some orientation to leading work on distributive justice, and outline a theory of justice that addresses the universally necessary and sufficient conditions for living well. It is only with reference to such conditions that the long-term preservation of opportunities to live well can be adequately conceptualized. In elaborating this theory, we address the current state of psychological research on what is and is not conducive to people living happily and explain its significance for the pursuit of affluence and economic growth. An important implication of this research is that material affluence is considerably less essential to personal happiness than it is widely assumed to be. A second important aspect of this research is that it provides a more informative basis for conceptualizing and pursuing the long-term preservation of opportunity than do measures of subjective well-being or conceptions of equal opportunity that are predicated on equal access to the occupational roles that are available in a society.[1]

Toward a Common Ethic of Sustainability

Across many fields of endeavor and inquiry, people are grappling with the ethical import of a trajectory of human existence that is trending ever more certainly toward catastrophe—catastrophe for humanity, catastrophe for millions of other species, catastrophe for the ecosystems on which we all depend.[2] Works exploring aspects of sustainability ethics and domains of ethics closely related to sustainability have begun to appear.[3] There are explorations in the ethics of sustainability, but these explorations have not yet coalesced into a well-defined domain of inquiry guided by a clear conception of what the domain is. We will offer a conception of the domain, addressing the scope, distinctiveness, basic principles, and virtues of sustainability ethics. With respect to principles, the philosophical task we will undertake is not to make moral law (whatever that would mean) but to articulate as clearly as possible the ways in which the basic commitments of common morality bear on matters of sustainability and entail specific responsibilities with respect to sustainability. The ethic of sustainability we present is thus an application of principles of common morality informed by the sustainability facts of life. Response to these facts is required by common morality, inasmuch as it regards us all as having a duty of care to make reasonable efforts to understand the circumstances in which we act and avoid doing harm. This is an approach sharply at odds with the common view that existing ethical frameworks are unhelpful in addressing matters of sustainability because they were created in and reflect a world of abundant frontiers. Many implications of the basic ethic we rely on have been ignored through long stretches of history, but they were not ignored by those who articulated this ethic most powerfully in the environmentally devastated world of ancient Greece to which we trace its origins.

Climate change and ecological overshoot are only the latest in a series of ethical crises that have inspired suggestions that a bold departure from common morality is required. These have included the Holocaust, recognition of systematic sexual discrimination, and hotly contested efforts to save species from extinction. All such crises warrant sustained moral self-examination, which may lead us to revise any number of beliefs, but there is no need to begin by assuming that the entire edifice of value and principle we have accepted is systematically corrupt and must be replaced. We already know a great deal about what is good for human beings and other living things. What we need is mostly an articulation of specific, action-guiding principles of ethics and political justice that reflect what we know.

We said in chapter 1 that the *preservation of opportunities to live well* is the normative focus of concern for sustainability. *Sustainability ethics* may consequently be defined as the domain of ethics and ethical inquiry pertaining to every sphere and aspect of human activity as they bear on the capacity of natural systems to provide opportunities at least as good as the ones available now indefinitely into the future. To say that sustainability ethics pertains to every *sphere* of human activity is to identify it as an aspect of universal personal ethics, social ethics, and political justice. *Personal ethics* pertains to everyone, everywhere, at all times, whatever they are doing. *Social ethics* concerns the norms of conduct of various institutions and forms of endeavor, and our roles within them, including professional roles. *Political justice* concerns the responsibilities of citizens and governments. Various roles, including those of professionals and of citizens, carry special responsibilities over and above those of common morality or personal ethics, as we use the term here. To say that sustainability ethics pertains to every *aspect* of human activity, as it bears on the preservation of opportunities to live well, is to imply that many aspects of our conduct make a difference to sustainability. It also assumes that factors as diverse as personal character, occupational role, institutional culture, and policy context shape our conduct. Personal virtues, basic commitments of common morality, principles derived from those basic commitments, codes of professional ethics, and ethical criticism of practices, institutions, and acts of government may all come into play.

Sustainability ethics is not confined to one sphere of activity, as business ethics is. Nor is it centered on concern with the value of the nonhuman natural world, as environmental ethics is. Sustainability ethics overlaps environmental ethics to the extent that the former concerns the life prospects of both humans and nonhumans, and it may identify principles or responsibilities that should supplement those presently acknowledged in business ethics or other domains of professional ethics. Sustainability ethics overlaps the domain of environmental justice inasmuch as sustainability is an issue of environmental justice, but not all issues of environmental justice are issues of sustainability. Environmental justice is concerned with fairness in the distribution of environmental burdens and benefits, but some (synchronic) unfairness in the distribution of environmental burdens and benefits may be consistent with sustainability, which concerns the (diachronic) preservation of opportunity into the future. Workers in solar panel factories might suffer unjust exposure to environmental toxins while performing work that contributes to a sustainable human footprint, for instance. Wider release of toxins could be a problem of both environmental justice and sustainability, of course, if it crosses a safe boundary for chemical pollution, but locally it might be a problem of environmental justice and not of sustainability.

Universal Respect for Persons

We begin from the idea of an ethic of respect for persons as rational beings. This ethic is often associated with Immanuel Kant, and it grounds moral duties of nonviolence, noncoercion, truth-telling, and mutual aid. The aim of Kant's *moral constructivism* is to show that rational beings have good reason to accept the authority of these duties. Suppose what we know in the abstract, or impartially, is that we are rational beings who have ends we wish to fulfill, that as finite beings we are vulnerable and have limited knowledge and capacity to fulfill those ends, and that there are a multitude of other finite beings whose aid could often enable us to better fulfill our ends at little cost to themselves. It would follow that it is rational for us all to adopt norms of respectful self-restraint and cooperation in sharing information and aiding each other when it is not too costly to do so.[4] In Kant's terms, we could not will to make it a universal law that beings such as ourselves would not cooperate in a variety of ways, if we realized that as finite rational beings we may have ends we will not be able to fulfill without the cooperation of others. Respectful self-restraint entails taking care not to cause harm, and fulfilling this *duty of care* requires attention to what one is doing and sometimes precautions such as investigating the qualities of one's conduct to be sure it is not harmful. This is an aspect of common morality long enshrined in English, American, and Canadian common law.[5]

Kant understands the duties of common morality justified in this way as authoritative in all spheres of human activity, and as such they constrain the manner in which citizens and states deal with each other. What is most obviously required is that such dealings and governing be transacted primarily through truthful and reasoned persuasion, instruction, and consultation—and only as a last resort through force or violence. These duties will also partially define the content of governance or law, such as through the enforcement of moral duties not to harm others. Yet natural moral duties are not enough to fully define the moral relations in which persons stand to one another. They do not fully determine property rights or the limits of acceptable imposition of risk and withholding of aid. The specification of such matters is an essential basis for cooperation, so individuals who cannot or do not choose to avoid interacting with one another, even if it is only through the remote effects of their conduct, have a duty to enter into a social contract that creates a common rule of law and governance.[6] In other words, if their actions have an impact on others' interests or life prospects, they are *obligated to negotiate fair terms of cooperation* governing their impact on one another.

The importance of this point can scarcely be overstated. Basic duties of interpersonal respect govern the context and character of negotiations and—where membership in a cooperative unit is involved—establish free and equal membership as the fundamental basis of cooperation under a common rule of law. Moral duties of mutual aid that are nonspecific as to the extent and timing of aid may also be given more specificity as a matter of law, so as to define civic duties of equitable shared sacrifice. For instance, a general moral duty to aid each other could be given specific content with respect to climate change as an allocation of emissions permits, fees, or something of the sort. Moral duties of care may also be codified and periodically updated as a matter of law, as patterns of harm that were once unexpected become better understood.

Basic aspects of this ethic of respect were already discernible in the philosophy of Socrates, who envisioned a social and political order regulated by reason-giving. Following Socrates, a central theme of Plato's and Aristotle's moral and political thought was that an ethic of respect for persons as rational beings requires that societies be regulated as much as possible by truthful and reasoned persuasion and instruction and only as a last resort by force and violence. Societies should bear a burden to make their laws and other norms of cooperation ones that reasonable people would have reason to accept, and they should also bear an inalienable responsibility to provide an education that equips all their members to understand the reasons and be rationally self-determining.[7] These are public responsibilities that apply in any context in which an ethic of respect for persons would entail a duty to enter into a social contract authorizing a common rule of law. Because understanding and the capacities of rational self-determination are only acquired with some assistance, the fulfillment of these educational responsibilities is essential not only to the legitimacy of the system of law but to its very enactment—to a rule of law becoming operative. The assumption that compliance with law is primarily a product of legal penalties is not supported by the evidence.[8]

This basic ethic of respect for persons and the concepts of sustainability introduced in chapter 1 are the basis of the principles that follow. Taking care not to harm others is a fundamental duty of common morality, and it would be deeply contrary to its spirit to regard acts that diminish opportunities to live well as anything but harmful to everyone's interests. In Kant's terms, we could not—as beings with the ability and need to cooperate—*will* to make it a universal law that we would live in a way that voluntarily destroys opportunities to live well in the future. Nor could we will to make it a universal law that we would fail to diligently investigate whether our actions may be destructive of opportunities, deceive others about the possible destructiveness of acts or practices, fail to negotiate fair terms of global cooperation when our lives are as globally entangled as they now are, or expose others to unreasonable risk of harm.

Principles of Sustainability Ethics

We have said that the totality of practices of a human collectivity is ecologically sustainable just in case it is compatible with the long-term stability of the natural systems on which the practices rely. Sustainability of this kind is basic, because it pertains to the preservation of the biocapacity or RNC that produces irreplaceable supporting, provisioning, and regulating "services" essential to human existence and well-being. A stable climate is essential to moderate reliable rainfall and other aspects of water availability, for instance, and there is no feasible replacement for naturally occurring fresh water on the scale of current human use. Our first principle of sustainability ethics concerns ecological sustainability.

1. *Take care to ensure that the totality of human practices is ecologically sustainable.*

This is a straightforward implication of a basic duty to take care not to harm, given the fundamental role of RNC in human existence and well-being. Interpreting and applying this principle requires measures of sustainability and clarity about how it should guide the conduct of different kinds of decisions and actors. Explorations in the ethics of sustainability have been largely confined to climate disruption and how to allocate a carbon emissions "budget" compatible with sustainability, but the planetary boundaries model described in chapter 1 offers a more comprehensive approach. Relying on this model, this first principle could be interpreted as an imperative to take care to ensure that none of the various planetary boundaries are crossed (e.g., with respect to biodiversity loss, phosphorus runoff, land cover conversion, atmospheric aerosol loading and nitrogen removal, pollution, etc.). With respect to the kinds of actors and decisions that might be involved, this principle identifies a responsibility of governments and government officials but also of individuals in their capacities as citizens and in their private and professional affairs to favor choices compatible with and conducive to the long-term stability of natural systems.

We have defined *throughput sustainability*, which is also important to the long-term preservation of opportunities to live well, in the following way: The totality of practices of some human collectivity is sustainable if and only if the material throughput on which it relies is compatible with the projected provisioning capacity of natural systems. Depletion of nonrenewable accumulated products of ecosystems (NNC), such as aquifers, is a less fundamental concern than ecological unsustainability so long as the total material throughput of human activities is less than a prudent maximum of RNC. If total human reliance on NC required only 30 percent of biocapacity, for instance, then ending the use of fossil water and fossil fuels might be manageable. Because it is presently about 150 percent of biocapacity and rising, humanity will find it impossible to sustain its current practices on the present scale. The economic throughput on which opportunities depend will decline, and the opportunities themselves will necessarily change. The opportunities will not necessarily be worse on the whole, but the prospect of declining throughput is nevertheless a serious threat to the preservation of opportunity. A fundamental duty of care not to harm thus entails a second principle of sustainability ethics.

2. *Take care to ensure that the throughput requirements of human practices are compatible with the projected provisioning capacity of natural systems.*

This too is a principle that identifies a responsibility of governments and government officials but also of individuals in their capacity as citizens and in their private and professional affairs. The application of this principle would involve the same kind of measures of ecological stability the first one does, insofar as it requires projections of future provisioning capacities of natural systems that may be overburdened, and it would require supply-side throughput measures as well. We will refer to these first and second principles as *principles of opportunity preservation.*[9]

An important implication of these first two principles being duties of care is that they have a precautionary aspect. Ecological and throughput sustainability require that natural limits not be exceeded, but *taking care* to ensure these limits are not exceeded would involve making investigations to thoroughly understand the circumstances of human actions and taking precautions to avoid doing harm, such as by holding in reserve some excess or redundant system capacity. It is also important to note that even when these aggregate limits are exceeded, the harm that can be anticipated is incremental. The endangering and diminution of opportunities to live well is not an all or nothing matter, so the underlying duty to take care to avoid harm would remain relevant. There is, in other words, no threshold effect that could justify saying it is "too late" to prevent disruptive planetary changes from occurring and we are all excused from making an effort to avert future misery.

A further important aspect of taking care to ensure that the totality of human practices is ecologically sustainable and sustainable on the supply side of throughput is that the relevant objects of choice include not only specific acts but—insofar as various actors have the power, authority, or influence to alter them—the attributes, norms, settings, structures, cultures, institutions, systems, and policies that are more or less conducive to ecological or throughput sustainability. This could be made explicit in a corollary to the first and second principles.

Corollary to 1 and 2: Take care to ensure that the human attributes, practices, institutions, systems, and policies within your control, authority, or influence are conducive to ecological and throughput sustainability.

The formulation *take care to ensure* could be glossed here, as in other contexts, as *make every reasonable effort to ensure,* where reasonableness must be judged in light of all that is at stake, the costs of exercising epistemic and preventive care, exerting influence, and so on.[10]

As individuals, our actions contribute to an unsustainable human footprint incrementally, globally, and often without any identifiable victim, making it hard to know what would count as full or sufficient compliance with these principles. The principles call upon everyone to make reasonable efforts, but how much is enough? How much difference will the voluntary actions of individuals even make? One answer is that *harm is incremental,* so our personal increments of pollution and waste do matter. Another answer, suggested in chapter 2, is that changing the way we individually live—and doing so voluntarily before adequate government policies are in place—is essential and can be efficacious in altering the political calculus of sustainability. It is nevertheless true that the translation of environmentally safe *collective* limits into publicly established and sanctioned limits on what *individuals* may do is the means by which we can ultimately determine what is and is not sufficient personal restraint. Our current situation with respect to sustainability is analogous to the early age of automobiles before there were speed limits and other traffic laws. A driver could slow down to reduce the risk of a collision, but how slow is safe enough? Democratically enacted speed limits and other traffic laws are the means by which we have collectively defined acceptable risk associated with automobile accidents. A carbon tax or permit (carbon cap and trade) system is one piece of what we need now to define the limits of acceptable environmental risk, and we need parallel policy interventions to fairly allocate the burdens of collective self-restraint with respect to other planetary boundaries. A system of user permits for NNC would also be required.

To say this is to affirm that even as environmentally conscientious individuals seek to live in ways that are compatible with sustainability, a fundamental burden we *all* bear—individual, institutional, and government actors alike—is to seek fair terms of cooperation in living sustainably. Earlier, we noted that seeking fair terms of cooperation is a basic requirement of interpersonal respect when individuals' acts have impacts on others, and it is undeniable that we are already interacting with others around the globe in ways that damage their interests. Our unsustainable energy and transportation systems already have damaging global reach, including an estimated 150,000 deaths each year attributable to climate destabilization.[11] We are having this impact while hesitating to negotiate a mutually agreeable climate treaty and other environmental protection treaties, and this hesitation is itself morally objectionable. It is a violation of the most basic norms of interpersonal respect to act in ways that impose risk and harm on others and to refuse to face them and negotiate reasonable limits on those impositions. Facing them as equals and specifying the details of our moral relations with each other in legal requirements is what is morally required. In practice, this implies that individuals have a responsibility to encourage government action to negotiate and enact binding agreements and to hold their elected officials accountable for failure to do so.

This leads to a third principle of sustainability ethics.

3. *Seek fair terms of cooperation conducive to sustainability. Actors whose actions affect each other have an obligation to cooperate in negotiating fair terms of cooperation in living in a manner that is collectively sustainable.*

Fair terms of cooperation would undoubtedly define not only the terms of participation in achieving a sustainable human footprint, but also what will constitute *wrongful* impositions of environmental risk on identifiable populations and jurisdictions. To that extent, the terms of cooperation in achieving sustainability also would address present matters of environmental justice, such as the issue of cross-border acid rain pollution that causes damage to forests in another country.

Fairness in cooperation requires that the terms of cooperation be known and clear to all parties and that acceptance of those terms not be predicated on one party being ignorant of what is known to another. People freely accept terms of agreement in the belief that doing so serves their prudential and moral interests, but they often lack information others possess, giving those others (if they are selfish) an incentive to misrepresent facts that might make agreements less attractive to the other parties. Clear and accurate disclosure of information is thus an aspect of fairness or mutual respect in civic cooperation and other voluntary transactions, such as commercial transactions. It is often termed *transparency* or, in the political sphere, a *publicity requirement*—a requirement to make certain things public, such as documents submitted as evidence or records of testimony.[12] Transparency in this sense is essential to the legitimacy of agreements, but it often falls short of ensuring that the agreements and commercial transactions people agree to actually serve their interests.

The *efficacy* or efficiency of civic and economic relations in serving people's interests requires a stronger form of transparency. It requires not only that all parties to the terms of cooperation or transaction know the relevant facts known to other parties, but also that *everyone knows what is actually at stake in the respects that matter to their legitimate prudential and moral interests*—in the respects that matter to what they do and should care about.[13] Achieving *full transactional transparency* of this kind requires coordinated efforts to discover and communicate the relevant facts, which may involve everything from the advancement of fundamental science, such as climatology, to the use of relevant sciences in investigating the probability of various costs and benefits of a proposed venture, as in preparing an environmental impact statement produced by an agency or business to justify a project that may have environmental risks.[14] There are reasons to consider the cooperative pursuit of public knowledge as a matter of fundamental constitutional concern—a matter we will discuss soon—but it is enough to see for now that in any case there will be public decisions to be made about the allocation of responsibilities to undertake and report investigations into matters of public interest, such as the environmental impact of business ventures. This is an aspect of transparency associated with the duty to seek fair terms of cooperation, as well as a basic moral duty to aid others when one can do so without much cost to oneself, a duty that we said should be given greater specificity as part of a society's terms of cooperation. The duties pertaining to this aspect of transparency, however, are an aspect of justice and not of common morality as such.

Nor are requirements to disclose the details of deliberations universal ethical duties. Transparency is widely regarded as essential to accountability and deterring corruption, but opposing arguments have been made in defense of secrecy as a condition of sound deliberation in various contexts. It is not clear that it always serves the purposes of sound judgment to require that all deliberations be made public, but how much secrecy is optimal is arguably a matter that should be publicly debated and determined.[15] The presumption against secrecy should be especially high when there is risk of corruption and serious harms that may not become apparent for many years, such as the health effects of some pollutants and climate impacts of current carbon emissions. However, these are again considerations of sound public policy and justice, not what is owed to others as requirements of basic moral respect.

This having been said, we can derive a fourth principle of sustainability ethics from a basic ethic of interpersonal respect.

4. *Do not obstruct transparency and cooperation with regard to sustainability.*

Obstruction of transparency could take, and has taken, a variety of forms, including corporate denials of matters of scientific consensus that the guilty firm's own scientists accept, creating a false appearance of scientific controversy, orchestrating campaigns to discredit honest scientists doing valuable work, misrepresenting the nature of science in the service of casting specious doubt on entire fields of science, and sowing confusion, all through corporate statements or proxies (sponsored websites, news outlets, industry front groups, and politicians; fake citizen groups; forged letters; scientists whose funding requires them to not reveal their funders; etc.).[16] Obstruction of cooperation may occur in any of these ways and more directly through lobbying; lawsuits; sponsoring politicians willing to cater to industry's short-term interests; and providing legislators with draft legislation that weakens environmental standards, enforcement, or both. Obstructing the cooperation in which others might engage in fulfillment of their own ethical obligations is an offense distinguishable from failing to fulfill one's own obligation to seek fair terms of cooperation. This is an aspect of the fourth principle that is not entailed by the third principle. It can nevertheless be defended in light of the third principle as an application of an implicit universal moral duty not to prevent others, by force or deception, from fulfilling their own moral obligations.

Obstruction of transparency and cooperation with respect to sustainability displays profound disrespect for individual self-determination and the collective self-determination of societies, and the evidence of massive obstruction of this kind is one reason to think the institutional structure and culture of the global economy is not conducive to sustainability. In creating the kinds of for-profit corporations that now exist, modern societies have conjured inventions of law that command unlimited financial and legal resources, and they have wagered that that they will be able to provide oversight and regulation sufficient to ensure that the activities of these corporations are consistent with the public interest. In essence, the fate of modern societies is predicated on maintaining the upper hand in an informational arms race, even as the pace of technological innovation quickens and corporate resources dedicated to private-interest science, public relations, advertising, litigation, and lobbying are allowed to grow without limit. A government doing its job to protect the public interest would reassess the wisdom of this wager and consider what modifications of corporate law are in order.[17] Societies and their governments have strong prudential and moral interests in creating and sustaining institutions and systems that are conducive to sustainability and transparency with respect to sustainability, and the implications of this fact for research, education, and other institutions of public knowledge are substantial. We will return to this topic ahead and in chapter 4.

Obstruction of transparency about the environmental impact of products and business practices is not only objectionable as dishonest and a barrier to seeking fair terms of cooperation in pursuit of sustainability, but also may induce hazardous reliance on vulnerable systems, exposing people to risk of harm they have not voluntarily accepted and might otherwise choose to avoid.

Inducing hazardous reliance is a way of exposing people to unreasonable risk of harm and thereby failing to take care not to harm them. Consider as an example that causing people to be at a party on a yacht at sea that cannot return them safely to port exposes them to unreasonable risk of harm and that it may occur in any number of ways—for example, by not paying attention to the yacht's safe occupancy limit and inviting people in excess of that limit, by setting sail despite warnings that violent storms are converging on its intended route, by using so much fuel to make ice for the champagne that it cannot return to port ahead of the storms, by leaving the life vests behind to make room for the champagne and failing to repair a leak in the fuel line, and so on. An aspect of the hazard associated with unsustainability is that excessive reliance on systems of limited capacity is a determinative causal factor in the growing probability and magnitude of harm likely to occur through ecological damage and depletion of natural capital.

The wrongness of putting people in harm's way by inducing or causing hazardous reliance can be captured in a fifth principle.

5. *Do not subject individuals or collectivities to detrimental reliance. Do not cause anyone to be in a position of fundamental reliance on hazardous or vulnerable systems or resources—systems or resources that cannot be relied on without exposure to unreasonable risk to their fundamental interests.*

This principle identifies imposition of risk per se as a form of wrong, and it focuses on the kinds of *systemic* risks that are at stake in discussions of sustainability: risks that ecosystems will collapse or that basic societal systems will suffer sharply declining capacity before sustainable alternatives can be developed and scaled up to replace them.[18] It addresses acts of *inducement*, in which people are induced to rely on something that is already hazardous or unreliable or will become so as a consequence of the reliance, and it addresses acts that cause something that is already fundamentally relied upon to become less reliable or adequate. An example of the former is the conduct of the fossil fuel industry in discouraging action to limit the use of their products. Any large-scale business that has the capacity to profitably develop and market products more conducive to global sustainability than its current products and fails to do so is arguably in violation of the principle of detrimental reliance. An example of the latter sort would be the poaching of fish in territorial waters, where islanders survive on fish caught by traditional methods within a few meters of shore.[19]

The principle may apply to the actions of specific individuals and collectively to the whole of a society or civilization. Regarding the latter, it says, in essence, that it is wrong to impose unreasonable risk on future generations by causing them to be in a position of fundamental dependence on systems that cannot be dependably relied on to provide them with adequate opportunities to live well. It is also designed to capture both the wrongness of inducing people to live unsustainably, putting them at risk, and the wrongness of putting future generations at risk through present unsustainable living. In general, it seems to capture what is morally troubling about unsustainability in a way other principles do not.

The principle that one should prevent harm that one can prevent at little cost to oneself has been invoked with good reason as justifying investments in reducing carbon emissions and stabilizing Earth's climate.[20] If there were no other reason for individuals and government and corporate actors to make the modest sacrifices that may be sufficient to avert catastrophic climate disruption, this principle would still provide one. However, because it applies independently of any causal contributions the actor may have made to the prospect of harm, it does not capture salient aspects of the wrongness of sustainability-related acts that put others at risk. One should rescue a drowning child who is at risk through no fault of one's own, but failing to do so is a very different matter from inviting a child to a birthday party big enough to capsize the ill-equipped yacht it takes place on and sailing off into storms.

The Dust Bowl of the 1930s provides an illustration of people being induced to live unsustainably and thereby subjected to unreasonable risk of harm. Homesteaders were induced to farm a region unsuitable for farming and largely destroyed the grasslands constituting North America's second-largest ecosystem.[21] Named the Great American Desert in 1820, the high plains grasslands later rebranded the Great Plains were designated by surveyors as too dry for farming. Nevertheless, with encouragement from the railroads and prairie state senators, the Enlarged Homestead Act of 1909 promoted dryland farming by distributing parcels of undeveloped federal lands. Then, as homesteading peaked in 1914 and World War I began, the US government encouraged planting of more wheat in response to the exclusion of Russian wheat from global markets. For a few unusually moist years, high plains wheat was profitable. Farmers expanded, taking on debt justified by a high price for wheat, and with the war's end and falling wheat prices they expanded again to cover their debt. At each step of the way, they plowed up the native grasses that anchored the soil and an ecosystem sustaining hundreds of species. In the drought of the 1930s that followed, the unanchored soil was gathered by winds into rolling mountains, ten-thousand-feet high or more, blinding and suffocating cattle, obliterating roads, and dropping thousands of tons of dust on cities hundreds of miles away. Infants died of "dust pneumonia" and birth rates plummeted. A quarter of a million people who had been induced to settle and farm a region that had never supported more than a few Native American hunting camps and villages fled, having lost everything, leaving behind one hundred million acres in ruin.

Virtues of Sustainability

We have counted virtues of character among the factors that may qualify as conducive to sustainability, and we have noted that *judgment*, a virtue of intellect essential to a virtuous state of character, comes into play in applying principles, including those framed previously. It is fitting that we should now say a few words about good judgment and character and their conduciveness to sustainability. In doing this we must point out that much of what has been written about virtue during the past thirty years has been shaped by the aspiration of *virtue ethics* theorists to show that excellent ethical decision-making does not, or need not, rely on principles or rules at all. We do not share this theoretical aspiration to establish virtue ethics as an alternative to other theories of moral decision-making or right action. Our position in a few words, all too metaphorically, is that principles and virtues need each other.[22]

Moral virtues are clusters of related dispositions of desire, emotion, perception, belief, conduct, and responsiveness to reasoning and evidence.

Virtues equip people to be appropriately responsive to the value of things and do what is reasonable. Individually, virtues pertain to recurring features of human existence such as encounters with danger (which call for courage) or difficulties in achieving a worthy goal (which call for perseverance and endurance). When they are present together in one person and mediated by good judgment, they form a virtuous state of character that equips the person to act well in a wide variety of circumstances. Considered as merely habitual and not mediated by good judgment, individual virtues might be blind to relevant factors and lead people to act badly—as a person may do in acting out of loyalty to a friend and with insufficient regard for other things of value. By contrast, a *true moral virtue* or virtuous state of character would be sensitive to the full range of ethically relevant factors in a situation, and it would be guided by good judgment. The moral motivation associated with true virtue is (ideally) responsive to everything of moral value that is at stake in the situation in which a person acts, and this is only possible if a person is well attuned to everything of moral relevance. The heart of moral motivation is valuing what has moral value, beginning with persons and other sentient beings, the goodness of their lives, and the necessities essential to their opportunities to live well. Moral motivation disposes a person to be appropriately responsive to the goodness of what is morally good and the badness of what is morally bad in the world she inhabits.

We have not identified principles of sustainability pertaining to the noninstrumental value of the natural world, nonhuman life forms, and what is good for such life forms, but being virtuously motivated would nevertheless involve being responsive to that value. The categories of value that can be usefully distinguished in this context are intrinsic, noninstrumental, and instrumental. We regard *intrinsic value* as pertaining to sapient beings and their inherently admirable excellences or virtues. *Noninstrumentally valuable* entities, activities, and qualities would be those valued for themselves by sapient beings, such as the rewarding activities of a good life and objects of attachment that are the foci of such activities.[23] *Instrumental value* could then be understood as dependent on a relationship to the first two forms of value. Aspects of nature may have noninstrumental value, insofar as they are valued noninstrumentally (e.g., loved or enjoyed) by humans or other sapient beings, and they may be valued instrumentally for their material contributions to a good life (e.g., as ecosystem services).

Good judgment is grounded in attunement and responsiveness to the value of things—in dispositions of desire, emotion, perception, and beliefs that are formed in the acquisition of moral virtues—but also in forms of understanding and the capacity and inclination to think things through. The good judgment that equips a person to act well in a complex world is impossible without both moral understanding and understanding the complex natural processes that connect human choices to outcomes in the world. An education that equips a person to understand how the world works is in this respect a natural culmination of the moral formation of a responsible person. The moral understanding essential to good judgment would be similarly useful in bringing matters of value into focus. The identification of patterns or *universals* is basic to understanding in both the natural and moral domains. In the latter, we have made a start toward identifying broad principles of sustainability ethics, and later in this chapter we will address some universal requirements of human well-being that will help us understand what is involved in preserving opportunities to live well.

In the meantime, we will note that there are individual virtues that should qualify as conducive to sustainability. Following Plato, we could begin with wisdom, justice, moderation, and courage. Under the heading of wisdom or good judgment, conduciveness to sustainability would require everyone to understand matters of sustainability and human well-being, distinguish important truths from inconsequential distractions and propaganda, critically assess prevailing norms and practices, assess risk accurately, and think creatively about how to live flourishing lives in ways consistent with sustainability. Under the heading of justice, sustainability would be facilitated by virtues of cooperative local and global citizenship, goodwill, and respect for others associated with the principles enumerated previously.[24] Under the heading of moderation, we would do well to cultivate endurance in resisting needless luxuries and inducements to measure success by conspicuous consumption, together with a steady regard for the true value of things. Finally, we must all find the courage to own up to the challenges at hand and do what is right in the face of hazards we would rather not contemplate or discuss.

These virtues are as applicable to individuals in their private affairs as in their professional and leadership roles. We should all display these virtues and nurture them in others. Yet recognizing this should not obscure the fact that progress can only begin with a few individuals who are willing to accept the burdens of "responsible initiative in a society as yet untransformed," and are able to display the virtues of such initiative: "a combination of knowledge to recognize urgency, the self-discipline to prioritize, the justice to care about contributing to the common good, and the sheer courage to step forward."[25] This is a rendering of the Platonic virtues as we need them to be if progress toward sustainability is to begin. "The need for initiative is dramatized, not replaced, by an analysis of the need for state action to address sustainability," writes political theorist Melissa Lane, for the state "really consists of a huge number of individual actors jostling and contesting to earn the right to act in the name of the state."[26]

Politics as an Art of Sustainability

If failing to live in a collectively sustainable way is likely to prove very costly, and the obstacles surveyed in chapter 2 make achievement of a sustainable collective footprint unlikely without coordinated and systematic public efforts, then coordinated public action is evidently essential. A just politics would display leadership in an art of sustainability, whereby current opportunities to live well are reconciled with the long-term preservation of such opportunities—leadership in advancing sustainability through moral clarity, public education campaigns and curricula, policy interventions, and investment. "The challenge of sustainability is one which requires both political art and philosophical insight to address, and no one could succeed in meeting that challenge without a knowledge of the scientific, but also the ethical demands which it makes," writes Lane.[27] Technology has a role to play—and there is no prospect of success in the pursuit of sustainability without major policy *and* technological breakthroughs—but technological advances alone are predictably outrun by failures of foresight, self-restraint, and cooperation. Cooperation requires leadership and shared understandings resting not just on individual enlightenment, but also on social and civic norms of conversation, attention, and response to problems and the troubling emotions they elicit.

There is historical precedent for regarding politics or statesmanship as an art of sustainability, an *art of preserving opportunity to live well*. This precedent was not set in the modern period of European political thought on which the American framers drew, however, but in the ancient Greek tradition from which much of their understanding of a constitutional government was ultimately derived. Writing in 1689, John Locke supported the enclosure of farm and pasture land in England that removed land from common use, creating a landless working class by 1820. He argued famously that God had bestowed the Earth on Man for his common use and enjoyment and that no portion of it could be justly enclosed by anyone for his exclusive use *unless* as much and as good remained for everyone else. He argued infamously that this "Lockean proviso" had not been violated in England because vast American continents remained available, neither owned nor enclosed by their native inhabitants.[28] With the conquest of these continents in progress, neither Locke nor any other European political theorist of the seventeenth or eighteenth century concerned himself with limits to "progress," even as the dispossessed were encountering harsh limits to their own life prospects. By contrast with Locke, Plato wrote his *Republic*, the first great work of political theory in the European tradition, at a time when Athens could no longer export its landless poor to colonies, which was its way of buying domestic tranquility through a kind of economic growth. The question just below the surface of Plato's theory of justice was how a society can be sociopolitically sustainable if endless growth is not an option.

Athens, Sparta, and Rome were unique to their world in escaping the social conflict and political instability that were otherwise pervasive, and the common explanatory factor was that they were uniquely successful conquest states.[29] Athens sent a large proportion of its poor over time to colonies established in conquered territories and used taxes imposed on subject populations to subsidize those who remained, but the Peloponnesian Wars with Sparta brought an end to this. It was in this context after the wars, and facing serious environmental problems, that Plato wrote the *Republic*. Philosophically, it is most obviously about the nature of justice or goodness and the relationship between goodness and happiness, but the background concerns that animate the dialogue revolve around *overconsumption*. The theme of excessive and ruinous consumption is invoked through the language of greed and injustice and more specifically of "unnecessary" and "lawless" desires.[30] The "true" and "healthy" city of book 2 is contrasted with the luxurious "city with a fever."[31] People in the former live simply, and everyone's needs are met even in old age; the city is sustainable across generations; it is a classless, unregulated partnership, with free and mutually advantageous exchange of goods; people enjoy sex but limit children to "no more than their resources allow"; and they thereby avoid both poverty and war.[32] This "healthy" city might be Plato's image of Eden. The unhealthy city is, expressly or by implication, none of these things. Its first unhealthy choice is to eat meat, which requires hunters, herds, *more land*, and doctors.[33] The addition of further luxuries requires more resources, even more land, and therefore an army and a policy of military expansionism. This is a portrait of Athens itself, and the story of the ensuing books of the *Republic* is one of restoring the luxurious city of Athens to a healthy state of equilibrium, a state of social harmony achieved through equity in the distribution of opportunities and products of social cooperation.

We know from other texts that Plato's concerns about sustainability and excessive consumption were not limited to the relationship between war-making and limited resources. By the time Plato was born, the landscape of Greece had already suffered from overgrazing by sheep and hillside deforestation, which had given rise to a number of legislative acts to replant hillsides and limit erosion.[34] Writing with a sound understanding of floods, erosion, and water scarcity caused by deforestation, Plato observes in the *Critias* that "Attica of today is like the skeleton revealed by a wasting disease, once all the rich topsoil has been eroded and only the thin body of the land remains."[35] He goes on to describe a city of Atlantis, densely populated, incessantly occupied with commerce, and ultimately ruined by intoxication with luxury and wealth.[36] The lesson, as in the *Republic*, is that when "possessions become pursued and honored," human beings exceed the limits of prudence and justice and come to a bad end.[37] The antidote he prescribes is a better understanding of what is essential to happiness and living well, an education foundational to happiness and justice, and a kind of justice that ensures no one lacks the means to provide for his own needs.

Athens was coming to grips with the hazards of luxury and economic expansion through military conquest, and the moral and political thought of Plato and his student, Aristotle, offer precedents for the work of sustainability theorists who are presently coming to grips with the hazards of a global economy that has been growing by $7 trillion per decade. There are echoes of Plato's concern with sustainability in Aristotle's works, especially in the latter's remarks about misguided accumulation of wealth, the importance of limiting population, equitable distribution of opportunities to live well, and injustice being the most important general cause of the collapse of regimes.[38] Aristotle's moral thought has been immensely influential in recent years, and it is an important point of departure for the theory of justice we outline later in this chapter. A central thrust of Aristotle's philosophy, like that of Socrates and Plato, was to undertake a reevaluation of values—especially materialism and status competition—that presented themselves as conducive to neither happiness nor a stable civic order. Psychologists today have the tools to investigate more rigorously than Aristotle could the relationships between materialism, social competitiveness, and happiness, and enough of his claims have been tested and vindicated that there is now a field of *eudaimonistic psychology* incorporating the Aristotelian idea that living well (*eudaimonia*) is largely a product of fulfilling human potential in positive ways.[39]

All of this suggests a two-fold political art of sustainability. Its first task would be to preserve the natural bases of opportunity so as to not exceed the throughput that can be sustained indefinitely. Managing this preservation requires knowledge of human motivation and what enables people to experience their lives as happy or going well, in addition to knowledge of natural systems. A second and related task would be to make the best use possible of the throughput that is consumed so as to provide the best opportunities for living well consistent with the five principles of sustainability ethics. Knowing what is and is not essential to people experiencing their lives as happy or going well is fundamental to success in this second task as well. Eudaimonistic psychology is foundational to both of these aspects of the art of preserving opportunity.

Background to a Theory of Justice

Kant's deduction of principles of universal morality depended on only the most general aspects of the "human condition" or situation of finite rational agents. Anglophone political philosophy since the 1960s has often relied on a related kind of *constructivism* to justify an impartial specification of just or fair terms of social cooperation. The work that set this in motion and has dominated discussion is John Rawls's 1971 classic, *A Theory of Justice*, and his subsequent refinements and extensions of the theory in *Political Liberalism* (1993), *Collected Papers* (1999), and *Justice as Fairness: A Restatement* (2001). Together, these works offer an impressively developed and defended vision of a just society in which government power is mediated by public reason-giving grounded in common constitutional values and norms of reasoning and evidence, a vision in which citizens are accorded the maximum feasible freedom to pursue their distinct conceptions of how to live a good life.[40] We will present Rawls's theory because it has dominated the landscape of political philosophy and because it is the most relevant point of departure for our own. We will borrow its constructivist methodology and endorsement of equal basic rights and liberties, but will in other respects use it as a foil for our own approach. We aim to do justice to matters of sustainability in ways Rawls's theory does not.

The theory begins with the idea of a society as a "fair system of cooperation over time from one generation to the next," the idea of citizens as "free and equal persons" who are open to finding and agreeing to fair terms of social cooperation, and the idea of principles of justice as a specification of those fair terms of cooperation.[41] Contemporary democracies exhibit a plurality of conceptions of a good life, and the liberties associated with free and equal citizenship protect this, as long as the conceptions are *reasonable*, in the sense of being compatible with fair terms of cooperation predicated on free and equal citizenship. In the circumstances of *reasonable pluralism*, the legitimacy of society's basic structure—the constitutional principles regulating its major institutions—will rest on an *overlapping consensus* in which people who subscribe to different conceptions of a good life nevertheless find diverse reasons of their own to endorse a basic structure.[42] A willingness to find fair terms of cooperation will ideally lead them to endorse common *principles of justice* and associated premises of *public reason*, which Rawls describes as a "shared basis for citizens to justify to one another their political judgments."[43]

The *original agreement* regarding constitutional principles would thus have two parts: basic principles of justice and "an agreement on the principles of reasoning and the rules of evidence in the light of which citizens are to decide whether the principles of justice apply, when and how far they are satisfied, and which laws and policies best fulfill them in existing social conditions."[44] These principles of reasoning and rules of evidence would be those of "common sense, and *the methods and conclusions of science*, when not controversial," and Rawls goes on to say that they apply not only in constitutional matters, but in legislation and public policy generally.[45] He insists, as other leading theorists of democracy have, that matters of consensus within relevant communities of scientists should be authoritative in judging empirical matters at stake in the public sphere.[46] The *duty of civility* citizens owe each other "requires us in due course to make our case for the legislation and public policies we support in terms of public reason" or by appealing to the values of a just constitution applied in light of common sense and scientific reasoning and evidence.[47] Some citizens might prefer to selectively exclude conclusions of contemporary science in light of their personal religious or philosophical worldviews, but justice would not permit this. The thrust of Rawls's argument is that the terms of public reason outlined are the only feasible and mutually respectful basis for cooperation in the conditions of contemporary democracies. Current standards of evidence in the sciences are the most neutral and authoritative basis for coming to agreement on matters of fact available, and there is no governing a society without coming to agreement on matters of fact germane to the public interest.

One reaction to this aspect of Rawls's theory is to agree that the idea of *public reason* is immensely important but object that the ideas of legitimacy and civility do not do it justice. Allen Buchanan has raised this objection and offered an alternative view that begins by observing how

limited we all are in what we could know without the *epistemic* or knowl-
edge-related cooperation of others. This leads him to focus not on norms
of public reason that govern citizens' political dialogue, as such, but on a
fundamental human need for "liberal institutions," defined as institutions
that reduce prudential and moral risk by sorting truth from error and
disseminating truth.[48] We all have fundamental interests, both prudential
and moral, that are served by truth, he argues; we need truth both to
advance our own well-being and to avoid doing harm to others. From
this perspective—which accords with our ideal of full transactional
transparency—we all have a prudential and moral interest in the existence
of institutions that respect norms of reasoning and evidence and serve the
public interest by disseminating the truths we most need to know in order
to live well. Buchanan has described these institutions as *veritistic* or truth-
telling, but they can also be described as institutions of *public knowledge*.[49]
Rawls developed and refined his theory of justice with a focus on some
important aspects of the human condition, but these did not include
human epistemic vulnerability and the importance of *epistemic goods*
(truth, knowledge, and understanding). His obvious concerns were equal
citizenship, inequality, and the significance of multicultural contexts for
constitutional legitimacy.

The *problem of distributive justice* is how the institutions of a society's
basic structure are regulated "so that a fair, efficient, and productive system
of social cooperation can be maintained over time, from one generation to
the next."[50] The task of Rawls's two principles of justice is to "assess the
basic structure according to how it regulates citizens' shares of primary
goods."[51] These *primary goods* are "all-purpose means" that citizens "need as
free and equal persons living a complete life," and the availability of primary goods to a person over the course of her life is equated with her life
prospects.[52] Rawls identifies five kinds of primary goods: (1) "basic rights
and liberties"; (2) "freedom of movement and free choice of occupation,
against a background of diverse opportunities"; (3) "powers and preroga-
tives of offices and positions"; (4) "income and wealth"; and (5) "the social
bases of self-respect."[53] *Wealth* is defined expansively as legal control over or
rights to use, receive, or benefit from "exchangeable means for satisfying
human needs and interests," including such things as "access to libraries,
museums, and other public facilities."[54] Wealth remains an economic category for Rawls, as one might expect, but access to epistemic and environmental goods might qualify as wealth under arrangements that make
them *exchangeable*. Rights of access to information obtained for a fee are

exchangeable, and rights to release carbon into the atmosphere would be exchangeable under a cap and trade system of emissions fees.

Rawls's defense of his principles is complex, but it relies on a memorable thought experiment, whereby the impartiality of representatives who are to determine the principles is assured by their ignorance. He asks us to imagine that these representatives are in what he calls an "original position" prior to the creation of a government, and that they know the general truths of human nature and society, including "the basis of social organization and the laws of human psychology," but they do not know particular facts about themselves, their place in society, or those they represent.[55] Rawls argues that from behind this "veil of ignorance" they will select principles of justice that guarantee equal and extensive rights and liberties, fair equality of opportunity, and limits on inequality:

(1) Each person has the same indefeasible claim to a fully adequate scheme of equal basic liberties, which scheme is compatible with the same scheme of liberties for all; and

(2) Social and economic inequalities are to satisfy two conditions: first, they are to be attached to offices and positions open to all under conditions of fair equality of opportunity; and second, they are to be to the greatest benefit of the least-advantaged members of society (the difference principle).[56]

Fair equality of opportunity "is said to require not merely that public offices and social positions be open [to all, in accordance with talent] in the formal sense, but that all should have a fair chance to attain them. ... Those who have the same level of [natural] talent and ability and the same willingness to use these gifts should have the same prospects of success regardless of their social class origin, the class into which they are born and develop until the age of reason."[57]

Some brief explanatory remarks are in order. A first observation is that neither the "fair value" of the political liberties guaranteed by the first principle, nor the fair equality of opportunity (FEO) guaranteed by the second principle are likely to be achievable without success in honoring the *difference principle* (DP), which "prevent[s] a small part of society from controlling the economy, and indirectly, political life as well."[58] FEO requires that children all have comparable chances to develop their talents and acquire the qualifications for desirable employment and offices, and this is impossible if some families control vast wealth while others are poor or destitute. However, even if FEO were achieved, it would not in itself ensure everyone adequate opportunities to live well. It ensures a fair competition for the available positions, but the qualities and rewards of those positions may be so skewed as to constitute a winner-take-all system—a system that gives only a lucky few the means to live well. The DP speaks to this problem as well. It offers wider access to more desirable opportunities by limiting and structuring unequal access to primary goods in ways that contribute as much as possible to the life prospects of those who are least advantaged by their social class origins. Rawls argues that a system structured to allow more inequality than this would not be just or compatible with free and equal citizenship.

We argued in chapter 1 that long-term preservation of opportunity is the normative heart of sustainability, and because FEO pertains to opportunity, it may seem most relevant to sustainability. It is, however, quite problematic in this regard. Rawls conceptualizes FEO in a way that depends on there being a common pool of positions and offices for which citizens from different social class origins will compete, but the pool is not fixed over time, making cross-generational comparisons infeasible. In actual societies—and especially ones in which educational systems are growing—the pool of positions on offer may continually change, ensuring that the terms of FEO could never be satisfied. If all social classes were equally well represented at all levels of an educational system at each stage of the system's expansion, then FEO might be achieved for a succession of birth cohorts. However, if some social classes trail others in their participation at ascending levels of the system, as they typically do, members of different classes may not face the same structure of opportunities when they reach a given level of educational attainment. They would not be competing for a common pool of positions and offices, and there may be no cumulative progress toward equality of opportunity between social classes. Although Rawls is explicit about his principles applying only within a generation and not across generations, the boundaries of a generation are ill-defined, and FEO could scarcely provide useful guidance for the pursuit of justice if the conditions for its application did not persist as long as it takes to implement an intervention designed to satisfy its requirements—a duration that likely would be measured in decades.

A Theory of Justice was published in an era in which educational institutions were widely regarded by sociologists as secondary institutions shaped by economic relations and without reciprocal influence of their own. It is likely that Rawls shared this view, for in societies that exhibit the class inequalities that concerned him, it is only on the assumption that schools are merely secondary institutions that they could play the role required by FEO, equitably transforming native talents into the bona fide job qualifications required for a common pool of positions of employment. If educational institutions are not merely secondary institutions and have a shaping impact on the nature of work and degree and terms of occupational stratification—if they alter the structure of opportunity, in short—then as members of trailing social classes reach different levels of an educational system they will face different structures of opportunity from the ones faced by those they trail, and there will never be a common pool of occupational opportunities they can compete for "under conditions of fair equality of opportunity."[59]

To see whether the opportunities to live well are stable and not diminishing over time, we would need an independent measure of how good the opportunities are in the aggregate at each point in the time series. Yet, as we noted, FEO does not regulate the *quality* of the opportunities on offer, only the fairness of the competition for them.[60] It does not give a principled formulation to the intuition that the long-term preservation of opportunities to live well is a fundamental matter of political justice.

The DP is also problematic as an instrument of intergenerational justice. Rawls insists that it applies only *within* a generation, and the intergenerational alternative he suggests is fraught with difficulty.[61] What holds between generations, he says, is a *principle of just savings*.[62] He says, in effect, that if the society does not yet satisfy the two principles of justice, then it is required to engage in real savings "to make possible the conditions needed to establish and to preserve a just basic structure over time." Once these conditions are achieved, "net real savings may fall to zero ... (real) capital accumulation may cease."[63] The society is to be a "fair system of cooperation over time from one generation to the next," so the question representatives are to ask themselves behind the veil of ignorance is "how much (what fraction of the social product) they are prepared to save at each level of wealth as society advances, should all previous generations have followed the same schedule [i.e., the same rule stating a fraction of social product to be saved at any given level of wealth]."[64] Rawls's principles require only enough real capital accumulation to "establish and preserve" just institutions and satisfy his principles of justice, not to allow future generations to be ever wealthier. For example, if it were scientifically well-established and knowable behind the veil of ignorance that poverty is a source of misery but per capita annual incomes higher than $29,000 yield no additional measurable benefits for justice or human happiness, then the just savings schedule could stipulate that real savings or capital accumulation would fall to zero once the society reaches that per capita income threshold.[65]

Rawls wrote with the presumption that societies become richer in manu-factured and human capital over time and with no acknowledgment that natural capital is threatened and must figure into any calculation of real savings or capital accumulation. He does not contemplate real savings fall-ing below zero, and it seems intuitively wrong to allow it to do so, but in scenarios like the present—in which the productive capacity of ecosystems is being diminished and stocks of nonrenewables, such as fossil hydrocar-bons and groundwater, are being drawn down—neither the DP nor the principle of just savings has luminously clear applications. Rawls evidently thought the DP would not be adopted as a fair principle of distribution across generations because it might constrain citizens in the future to be no better off than those who were least advantaged in the past (because allo-cating more to those in the future could have no benefit for those least advantaged in the past), whereas people—both generally and as representa-tives behind the veil of ignorance—might prefer progress from one genera-tion to the next. By contrast, one might think that in a scenario like the present, in which those born in the future are likely to be a disadvan-taged class by comparison with those presently alive, the inequalities between present and future cohorts would be unjust under the DP, given how detrimental to future life prospects the excesses of present opulence are likely to be. Inequality across time would not be to the advantage of those who will be the least advantaged. Nevertheless, the spirit of the DP is to limit inequality in a cooperative society to the inequality that is most favorable to the least well-off by birth, and this is not generally applicable across generations.

The principle of just savings is conceived with cross-generational appli-cations in mind, but its application grows murky in contexts of declining natural capital. Rawls writes as if a negative real savings rate would never be acceptable, especially if it threatened the stability of just institutions, but he offers little explanation. Could such a savings rate never be acceptable at *any* level of wealth? Would it always be ruled out because it would always, at any level of wealth, threaten the preservation of a society's just basic structure? How would the balance of natural and nonnatural capital figure into calculations of just savings?[66] Once the relevance of natural capital is introduced, and with it the idea of natural limits to growth, it can no longer make sense to determine a just savings rate as a "fraction of social product to be saved." As a first approximation, a just savings rate would be a fraction of what is produced (manufactured and human capital) and a fraction of natural capital that *could* be sustainably withdrawn from natural systems but is not. Savings with respect to natural capital would take the form of natural capital withheld from use in order to enhance biocapacity. Would a reasonable application of the principle allow withdrawals of natural capital to rise, accommodating a growing population at a higher standard of living, until a just basic structure is established or human demands reach 80 percent of what is sustainable (leaving 20 percent unused as a kind of insurance), whichever comes first? If Rawls were alive and willing to follow us this far, he would presumably say that establishing and preserving a just basic structure is nonnegotiable and that the effect of the DP (in its *intra*-generational application) in combination with just savings would be to restrain both consumption and family size, even at the upper end of the income distribution.

It bears emphasizing that the foregoing features of Rawls's theory all pertain to justice *within* countries, and that theorists' conceptualizations of global justice were limited to matters of war and peace until recently. Rawls's own theory of global justice would not take us far toward formulating fair terms of global cooperation for stabilizing the climate and oceans, sharing the costs of adaptation, and so on. Having begun from his theory of justice for societies with ideals like those of the United States, he suggests a second original position contract, not between countries or states but between *peoples* who are prepared to accord each other respect as equals.[67] Other theorists of global justice have suggested instead a "one-shot" global contract that yields a global version of Rawls's socioeconomic egalitarianism or a less egalitarian system of universal human rights, or they have grounded human rights approaches in some other way.[68]

Our concern is to understand in general terms what might constitute a just basis for cooperation to address problems of sustainability. Any such basis has both global and domestic aspects, because the terms of global cooperation that might be agreed upon must also be implemented in domestic policies in the various participating countries. Rawls's theory of justice provides more help with the domestic sphere than the global sphere, but we will need a conceptualization of justice more suitable to understanding the long-term preservation of opportunity. We will also need a conceptualization of fairness in global cooperation that is specific to matters of sustainability.

Outline of a Theory of Justice

Our approach in this section will be to sketch a theory that identifies the basic function of governments and leading institutions as enabling all the members of a society to live well without regard to when they are born. The Greek term *eudaimonia* signifies living well or flourishing, and the approach is a kind of *eudaimonic constructivism* or fusion of Aristotle and Rawls.

The Eudaimonic Principle

We begin by taking a step back from the question Rawls puts to the representatives behind the veil of ignorance concerning the principles regulating society's major institutions and ask what the aims of these institutions should be. Our conception of these aims, inspired by the idea of *eudaimonia* or living well in Aristotle's ethical works, begins with the idea that we all want to live well—to live in ways that are good or admirable and that we experience as good or satisfying. Because people do all want to live well, despite having different ideas about what living well entails, representatives behind the veil of ignorance would converge on a fundamental constitutional principle of liberal or freedom-respecting eudaimonism:

Eudaimonic Principle: The institutions of a society exist *to enable all of its members to live well* and should provide opportunities *sufficient* to enable all to do so *and thereby provide each other* such opportunities.

We will assume that representatives would also begin by endorsing Rawls's first principle of justice and welcoming the democratic institutions it requires, and we will assume that the final clause of the eudaimonic principle entails a form of property-owning democracy, as in Rawls. We intend sufficiency of opportunities to entail a universal distribution of adequate real possibilities, and thus a kind of equal opportunity, but we will defer the details of this to chapter 4. When representatives behind the veil of ignorance consider what else institutions must provide in order for everyone to live well, they have available to them knowledge of human psychology. Aristotle argued that while people all want to live well, many are ignorant of necessities for living well entailed by human nature, and knowledge of these necessities should guide the design of institutions. We agree and see a basis for reconciling what is valuable in his ethics with the ideals of a pluralistic constitutional democracy.

What Human Beings Need to Live Well

It is now well confirmed cross-culturally and across the life span that there are universal basic psychological needs, the satisfaction of which is both essential to living well and strongly influenced by the way institutions function.[69] As it happens, these needs also play an important role in human motivation, regulating not only the direct impact of institutional designs on human well-being but also the impact of those designs on the quality of individuals' contributions to the fulfillment of institutional purposes. A workplace designed to be good for workers by facilitating the satisfaction of these needs could anticipate better worker productivity than a workplace that is not good for workers. This is one illustration of a broader implication of the research on basic psychological needs—namely, that well-being or happiness is psychologically associated with fulfilling basic human potentials in admirable ways.[70] The objective and subjective aspects of living well are significantly related to one another through the satisfaction of basic psychological needs, in other words.

Basic psychological needs theory (BPNT) posits three innate, *universal psychological needs* associated with the fulfillment of human potentials: the need for *competence* or efficacy, the need for *self-determination* or *autonomy* rooted in values with which the agent wholeheartedly identifies, and the need for *relatedness* or the experience of mutually affirming relationships. Frustration of these needs to fulfill personal potential is manifested in depressed affect, deficits of energy and experience of meaning in engaging tasks, elevated error rates, and symptoms of stress and psychic conflict, such as headaches and sleep disturbances. The role of standards of success or excellence in "positive" fulfillment of potential has been implicit and only recently addressed explicitly in the language of virtue in the larger theory of which BPNT is a part—namely, self-determination theory (SDT)— and in the positive psychology movement that has drawn on SDT. The experience of *competence* will not reliably occur without efforts that are confirmed by others as measuring up to standards of competence or excellence, and the experience of success in *self-determination* typically requires the possession and exercise of intellectual virtues, but it is in the satisfaction of *relatedness* needs that the role of virtues of character is clearest. A person's psychic well-being is impaired not only by the failure of others to affirm her worth, but also by her own failures to affirm the value of others. In some studies, the observed psychic benefits of offering unreciprocated altruism were even larger than the benefits of being a recipient of such altruism.[71] The importance of possessing virtue to being happy has been substantially vindicated.

The evidence of hundreds of studies spanning diverse cultures and life stages suggests that the satisfaction of *all three* of these basic psychological needs is essential to and a predictor of the experience of happiness and related aspects of psychological well-being. Of the three needs, autonomy is the one most likely to be challenged as a culturally specific Western value and not a universal need. Cross-cultural studies do not bear this out,

however. They suggest that the need to be self-determining in one's activities is no less acutely experienced in cultures that regard autonomy as an
alien value; restrictive and hierarchical cultures often do provide spheres of
meaningful self-determination that meet this need, such as forms of spiri-
tuality that offer scope for personal choice and expression.[72] Those inclined
not to cultural determinism but to economic and behavioral models will be
deeply suspicious of claims that altruistic conduct engenders happiness,
and they may insist that human beings are selfishly calculating in all their
decisions and conduct. The popularity and influence of these perspectives
notwithstanding, they are supported neither by evolutionary science nor
by experimental evidence.[73] Unselfish behavior is common, and research in
experimental gaming demonstrates "not only that people routinely indulge
in unselfish behavior, but also that people know that *other* people routinely
indulge in unselfish behavior."[74] Research in the SDT paradigm, meanwhile,
has shown through a variety of research methodologies that altruistic conduct is not motivated by "hedonic outcomes," which is to say a selfish
desire to experience the *pleasure* that people often experience in helping
others. In these studies, "the relation between helping and wellness outcomes [for the helper] was fully mediated by need satisfaction. That is,
when people helped others by choice, they, in doing so, experienced autonomy, competence, and relatedness, factors that in turn accounted for the
positive outcomes."[75] Humans evidently evolved to find inherent satisfaction in treating other people well and to experience less well-being when
they do not treat others well.

Subjective preference theorists may argue along similar lines that people
simply maximize the satisfaction of whatever preferences they may have
and that happiness is a function of preferences being satisfied. Because
wealth is useful in satisfying a wide array of preferences, such theorists
might then argue that a society with more wealth is necessarily a happier
society and that the responsibilities of governments with regard to happiness promotion can be distilled down to maximizing the society's aggregate
wealth. To the extent that this view attributes fundamental moral value to
preference satisfaction as such, to all preferences equally, and to nothing
else, it is a puzzling form of value subjectivism. If attributions of value really
were that subjective, why stop there? If we do not agree that persons have
value in their own right (i.e., intrinsic value), why would we think there is
any value in their preferences being satisfied? Yet, if we do agree that persons have intrinsic value, how could we attribute as much value to the
satisfaction of, for example, a preference to see people suffer as to a preference to serve society through meaningful work? Knowing that there are

biological facts about human wellness, how could we imagine it makes sense to make no moral distinction between Person A's preference to smoke cigarettes at will and Person B's preference to not have her health impaired and life shortened by secondhand smoke? Knowing that there are biological facts about human wellness and intimately related psychological facts about human motivation, vitality, and psychological needs, our view is that there is more than enough factual basis in human nature to underwrite a morally significant distinction between what people *need* in order to live well and what they merely *want* or prefer.

Returning to the alleged relationships between wealth, preference satisfaction, and happiness, there are several important bodies of contrary evidence. Wealth does indeed enable people to satisfy some kinds of preferences, but the satisfaction of such preferences beyond necessities of life or escaping poverty is not strongly related to happiness. Average life satisfaction in relatively affluent societies has not risen as such societies have become more affluent in recent decades.[76] Materialistic (extrinsic) life goal orientations deflect people from the satisfaction of their relatedness, autonomy, and competence needs, making them generally less happy and less successful in achieving happiness as they succeed in their life goals than people who have comparable success in attaining nonmaterial (intrinsic) aspirations, such as for personal growth, close relationships, health, community service, and the like.[77] The relationship between materialism and unhappiness has been shown to be not merely correlational but causal and bidirectional: Materialism causes people to be less happy, and unhappiness causes people to be more materialistic, grasping, and selfish.[78] People could just happen to be wealthy and have nonmaterialistic values that lead them to live in ways that actually make them happy, of course, but the good news for sustainability is that the evidence regarding wealth shows that—beyond meeting necessities and giving us the wherewithal to engage in activities that fulfill our potential in positive ways—it is not essential to happiness or living well.

The robust confirmation of BPNT would allow representatives behind the veil of ignorance to conclude that fulfillments of basic human potentials in ways that satisfy related psychological needs for self-determination, mutually affirming relatedness, and competence are *essential goods* that are both sufficient for living well and compatible with a plurality of reasonable conceptions of a good life. Although there are aspects of cultures and institutions that hinder human flourishing and sustainability, the diversity of forms of competence, mutually affirming relationships, and spheres of self-determination assures a wide scope for individual and cultural expression compatible with a public responsibility to ensure universal access to these essential goods. What can be impartially agreed upon is that living well requires the acquisition and exercise of virtues, development and use of powers or capabilities to act (hence opportunities and the resources to act), and understanding what one is doing. The necessities for living well embodied in persons will consequently take the form of *virtues* of character and intellect, *capabilities*, and *understanding*.[79]

This identification of virtues, capabilities, and understanding as developmental necessities for acting and living well is a reflection of the structure of human action. No life can qualify as well-lived unless it is successful in relevant respects, and a person's success in acting requires that she be well-equipped and disposed with respect to *all* of the fundamental dimensions of agency. There are three broad categories of potentials fundamental to action—intellectual, social, and creative—and all three must be fulfilled substantially in order for a person to act in ways that are both admirable and satisfy psychological needs for self-determination, relatedness, and competence. One's intellectual and social potential must be competently fulfilled in determining what to do in light of an accurate *understanding* of the world and in accordance with the valuing of persons and what is good for them characteristic of a *virtuous* state of character, and one must then deploy (social, intellectual, and creative) *capabilities* competently, in order to act well and satisfy one's needs for self-determination, mutually affirming relatedness, and competence.

Institutional Bases of Opportunity

Recognizing that there are necessities for living well that people can only obtain in favorable circumstances, representatives behind the veil of ignorance would agree that the arrangements and institutions of society should in general facilitate both the *acquisition* of personal qualities essential to living well and the *expression* of those qualities in activity constitutive of living well. They would agree that there should be major institutions whose functions are to promote the acquisition of the *internal* attributes necessary for living well and provide *external* contexts in which these attributes can be expressed in admirable and rewarding activity constitutive of living well. These institutions would include educational institutions, the basic function of which is to promote forms of development conducive to living well; epistemic institutions, the basic functions of which are to produce public knowledge and promote full transactional transparency; and places of work and leisure that provide contexts in which the need for competence can be satisfied in good work and other forms of admirable and satisfying activity.

Societies must also provide for other basic needs, and the functions of workplaces would also clearly include both producing and conserving other prerequisites for the activities of good lives. This is an important aspect of a just society being cooperative in such a way as to provide everyone real opportunities (i.e., all of the necessities) for engaging in the activities of a good life. The fundamental rationale for social cooperation is to secure necessities for living well that people cannot secure on their own, and our list of three internal necessities for living well—virtues, understanding, and capabilities—subsumes the fundamental roles for epistemic and educational institutions defended by Buchanan and Aristotle.[80]

Our principles of sustainability ethics require societies to preserve the ecological basis for opportunities to live well and not exceed a sustainable level of economic throughput. We can add now that so far as these natural bases of living well are concerned, a comprehensive measure of per capita natural capital will serve as a reasonable proxy for whether one generation is living in a manner consistent with the long-term preservation of opportunities to live well.[81] The preservation of opportunities to live well also requires sociopolitical sustainability, and Rawls provides an influential but contested theory of some aspects of such sustainability in his theory of justice and of savings (capital accumulation) adequate to securing and preserving just institutions. What his theory does not provide is a conception or measure of quality of opportunities that will permit comparisons over long expanses of time and across geographically and culturally distant contexts. His measure of synchronic equality of opportunity—defined in terms of the probability of members of different social classes obtaining desired occupations from a common pool of offerings—will not suffice.

By contrast, our conceptualization of universal necessities for a good life offers a suitably timeless account of what people need in order to live well and thus what is institutionally essential to the long-term preservation of opportunities to live well. It identifies the potentials we need to fulfill, the related psychological needs we must satisfy, and the kinds of activities we need to engage in to experience our lives as going well—as lives that are personally fulfilling and worthy of admiration. In doing so, the theory clarifies why escaping poverty is good but pursuing wealth without end is not positively related to happiness. As a measure of just institutions, it aims to balance efficiencies of production and service, on which current policy fixates, with an equally fundamental focus on the qualities of the experiences the institutions afford—their roles in providing contexts in which to fulfill human potentials in admirable and rewarding ways. Grounded in a large and growing body of internationally replicated psychological research, this conception of just institutions and the long-term preservation of institutional bases of good opportunities should be less controversial than a version of Rawls's theory that found a way to project variants of his principles of justice across generations, supposing this were possible.

The long-term preservation of institutional bases for living well implies sociopolitical sustainability, and it is reasonable to ask what forms of justice may be essential to sociopolitical sustainability and preserving these institutional bases for living well. The answer we can tentatively offer here is that the forms of justice essential to sociopolitical stability would include the equal basic liberties guaranteed by Rawls's first principle and a counterpart of his (synchronically applied) principle of fair equality of opportunity (the framing of which we will delay until chapter 4). Sociopolitical sustainability requires a high level of voluntary cooperation with the rules on which the society operates, and such cooperation is unlikely if those rules do not offer widespread opportunities to satisfy basic needs and live well. How much inequality societies can bear without collapsing is too big a question to resolve here, but there should be little doubt that the current extremities of opulence are indulgences that humanity can scarcely afford. The institutional reforms implied by our approach would to some extent and in some ways restrict socioeconomic inequality and inequalities of citizenship.

Global Cooperation Conducive to Sustainability

Having addressed at some length these matters of domestic intergenerational justice germane to the pursuit of sustainability, we will close by recalling that our third principle of sustainability ethics requires us and our governments to seek fair terms of cooperation conducive to sustainability. No theory can fully specify such terms of cooperation in advance, but the ethic of sustainability and theory of justice outlined in this chapter have implications for what would qualify as fair. These implications can be roughly formulated as four provisions:

1. The terms of cooperation in pursuit of sustainability should accommodate and encourage reforms and development consistent with the five principles of sustainability ethics and necessary to countries' satisfying the eudaimonic principle. Countries would be right to reject

as unfair any terms of agreement that would offer their citizens no prospect of living well.

2. These terms of cooperation should allocate emissions limits that deviate from a global uniform per capita basis to the extent required to satisfy provision 1.[82]

3. Because an allocated cap on carbon emissions only addresses climate change and nine other planetary boundaries have been identified as fundamental to ecological sustainability, allocated caps governing cooperation with respect to each of those boundaries should be negotiated, using provision 2 as the default model.

4. Where regional transboundary allocations of NNC must be negotiated and have implications for basic human rights or countries' progress in satisfying the eudaimonic principle (as in regional disputes over water), allocations should honor those rights and give priority to the fulfillment of basic eudaimonic needs. Allocations of water should be consistent with the basic right of access to clean water ratified by the United Nations General Assembly and affirmed by the UN Human Rights Council.[83]

Conclusion

This chapter aimed to define a basic ethic of sustainability and outline a theory of justice that can adequately conceptualize the long-term preservation of opportunity to live well and inform our understanding of fairness in allocating the domestic and global burdens of cooperation in pursuit of sustainability. Clarity about what we do and do not need to live well is immensely important to living sustainably, and we have outlined a theory of justice focused on what human beings need that is grounded in a robust body of research consistent with cultural traditions across the world but sharply at odds with the behavioral model underlying conventional economic and policy perspectives.

In the next chapter, we will elaborate on our brief remarks about schools as formative institutions, workplaces as institutional settings in which personal qualities are expressed in activity that is more or less characteristic of living well, and epistemic institutions as truth-discerning and disseminating institutions. Having addressed these details of institutional contributions to living well, we will discuss the growth dynamics of sociopolitical complexity and collapse. The growth of complexity is a matter of great consequence for sustainability and the preservation of opportunity over time. Societies tend to become more complex over time, and this growing complexity alters the structure of opportunity. It creates new and challenging opportunities, but also greater social and economic stratification and growing energy, material, education, and coordination costs. The collective marginal return on investments in complexity also generally declines, even as the costs of additional complexity grow. The growth of complexity alters the array of opportunities available in the society and in doing so challenges conventional thinking about the promotion of equal opportunity and reinforces our own view that an adequate conceptualization of sustainability or the long-term preservation of opportunity to live well requires a suitable account of what is inherently essential to living well.

Limits to growth are an inevitable aspect of discussions of sustainability, but the systemic relationships between economic and educational system growth are scarcely addressed, even as many call for education in sustainability. We will address this omission by discussing the interface of school and work in the context of a revealing model of sociopolitical complexity and collapse, contrasting the US system of *market credentialism* with a more equitable alternative. The resulting account of socioeconomic-educational complexity provides a diagnostic basis for proposals that might contribute to the long-term preservation of opportunity and reverse what is arguably a broad pattern of declining marginal collective benefit from investments in education and research.

4 Complexity and the Structure of Opportunity

The survival of just institutions is important to sustainability in a context such as the present, in which the aggregate of human activities exceeds what is ecologically sustainable. Without such institutions and the assurance of adequate opportunity for everyone to live well now, present necessities will predictably eclipse efforts to protect the quality and abundance of opportunities in the future. Therefore, it is important that we have a clearer conception of the functioning of just institutions and of how our present institutions do function, in order to understand better what kinds of reforms might enable institutions to provide opportunities to live well both now and in the future.

This chapter will elaborate on our brief remarks in chapter 3 concerning the nature of just institutions: epistemic institutions, the function of which is to identify and disseminate truths important to the public interest; schools as formative institutions; and occupational settings in which personal qualities can be expressed in activity that is more or less characteristic of living well. With a more substantial account of the just functioning of these institutions in hand, we will address the growth dynamics of sociopolitical complexity and collapse.

As we noted at the conclusion of chapter 3, the growth of complexity is a matter of great consequence for sustainability and the long-term preservation of opportunity. It alters the structure of opportunity, entails rising energy consumption and other costs, and exhibits declining marginal social return. Educational institutions facilitate the growth of complexity, altering the nature of work and the structure of opportunity, but conventional thinking about equal opportunity has assumed that schools are merely "secondary" institutions shaped by economic relations and without reciprocal influence of their own. If schools were merely secondary institutions, then they might be able to play the role required by the ideal of fair equality of opportunity—equitably transforming native talents into the bona fide job qualifications antecedently required for a common pool of positions of employment. Ironically, the character of attempts to promote equal opportunity in the United States has contributed substantially to the power with which educational institutions have facilitated the evolution of a less equitable structure of opportunity. Our diagnosis of this pattern and discussion of how things might be different sets the stage for our alternative to Rawls's principle of FEO, for which we issued a promissory note at the conclusion of chapter 3.

Having offered a broad conceptualization of the dynamics and costs of complexity, we will conclude the chapter with some proposals that might contribute to the long-term preservation of opportunity and reverse what is arguably a broad pattern of declining marginal collective benefit from investments in education and research. Universities are both educational and epistemic institutions and they play an integral role in the growth and unsustainability of the larger socioeconomic system. There are ways in which they could be more conducive to sustainability in their formative and informative aspects, and there are ways in which the larger educational system of which they are a part could have a more favorable impact on the shape of opportunity.

Just Institutions

We argued in chapter 3 that a just constitutional system would establish institutions and principles of cooperation designed to enable the cooperating members of a society to live well, paying particular attention to what individual persons cannot secure through their own efforts. Just constitutions would provide not only for the *internal* or developmental necessities for living well (which we identified as understanding, capabilities, and virtues), but also necessities for living well pertaining to the contexts in which individuals act and other *external* prerequisites of admirable and rewarding activity. We identified epistemic institutions, educational institutions, and workplaces as three (nonexclusive) categories of basic institutions important to the fulfillment of human potential in admirable attributes and the rewarding expression of those attributes in admirable activity. The task of this section is to fill out this conception of just institutions in somewhat more detail.

Epistemic Institutions

From an impartial point of view, the contemplation of human epistemic limitations commends norms of truth telling, information sharing, cooperative inquiry in the public interest, and institutions such as national academies of science that are designed to authoritatively present matters of scientific consensus for public consideration and policy. Put somewhat differently, it seems evident that there is potentially great advantage in making it a matter of fundamental constitutional concern to establish ground rules favoring "veritistic" institutions that seek and disseminate truth and understanding that is important to the public interest.[1] We touched on some of the advantages of such ground-rules in chapter 3 when we introduced the idea of full transactional transparency and outlined a theory of justice that identifies understanding as a necessity for living well. Our understanding of the world and the specific contexts in which we act is fundamental to success in our acts and endeavors—and therefore success in living well generally. *Voluntary transactions* are simply one category of acts of which this is true, and *full transactional transparency* is the term we have chosen to designate the information or understanding essential to the efficacy or efficiency of commercial, civic, and other voluntary transactions in advancing the legitimate interests of those who engage in them.

Beyond these basic points, we noted in chapter 2 that institutions of public knowledge involve a division of epistemic labor, whereby there are expert producers of knowledge and nonexpert consumers of it, and that the efficient flow of knowledge and understanding requires that *trust* be well-placed. Because we must all accept most of what we know or believe on authority, it is important that epistemic institutions be designed in ways that allow nonexperts to reliably judge who is epistemically trustworthy on what topics. One aspect of this is to require full disclosure of the sponsorship of research and the methodologies used, so that conflicts of interest can be assessed and independent experts can critique the reasoning and evidence on which claims are based. What else is required?

Rawls defines a *well-ordered society* as one "effectively regulated by a public conception of justice," where this requires that "everyone accepts, and knows that everyone else accepts, the very same ... conception of justice [and] principles of justice," and the society's basic institutions are "publicly known, or with good reason believed, to satisfy those principles of justice."[2] Noting that Rawls includes the epistemic norms that regulate *public reason* as constitutional principles on a par with principles of justice, it would be within the spirit of his theory to identify an *epistemic* form of constitutional well-orderedness consisting of universal acceptance and mutual recognition of acceptance of these norms of public reason. His theory would restrict this to the sphere of civic transactions where norms of public reason apply, but our own approach would generalize the relevant epistemic norms to voluntary transactions generally. We define an *epistemically well-ordered* society as a society in which "everyone accepts, and knows that everyone else accepts" the same norms of reasoning and evidence as authoritative, and the society's institutions of public knowledge are "publicly known, or with good reason believed, to satisfy" those norms of reasoning and evidence. Institutions of public knowledge would include all institutions that a just society would designate and rely on as sources of knowledge and understanding, and the authoritative norms of reasoning and evidence would be those of the communities of experts within whose purview a question falls. This is an ideal not easily satisfied, and it is only approachable if citizens are educated enough to have some understanding of the distinguishing features of serious inquiry, if institutions that profess to enlighten the public and their leaders (from mass media to think tanks, research institutes, and universities) honor relevant norms of serious inquiry, and if their honoring of those norms is adequately disclosed to trusted and trustworthy representatives of the public who possess sufficient expertise to judge the evidence disclosed.

Saying only this much leaves many questions unresolved. Are businesses that represent themselves as advancing the frontiers of energy, food, or communication technologies in the public interest epistemic institutions, subject to the requirements of disclosure and public scrutiny this suggests? From the perspective of the theory of justice we have sketched, the answer is that they *are* institutions of public knowledge subject to these requirements *if* an impartial judgment of the society's collective interest is that they should be. From a legislative standpoint, it is hard to see how—in the face of unfolding sustainability crises—we could rationally shield as much proprietary information in the interest of competitive advantage and new products as we now do, as if the abundance of new products should be among the paramount concerns of this historical moment. Consider, for example, that with the long-term security of groundwater at stake, it can hardly make sense to shield from public scrutiny the chemical content of the fluid used in the fracturing of shale formations to release natural gas.

Three further points are in order, all pertaining to the internal functioning of institutions of public knowledge. The first is that we have spoken of authoritative norms of reasoning and standards of evidence as those of the various authoritative communities of experts, but important questions often do not fit neatly within the confines of any one science or discipline. Institutions that sponsor work across established disciplinary boundaries and forms of expertise are thus essential to the project of public knowledge, even if the epistemic standards for synthetic investigations are a collective work in progress. Common constructs or bridging concepts must be invented and anchored (operationalized) in serviceable forms of measurement, innovative research designs must be developed, and conventions of disciplinary boundaries and cultures must be overcome. The structures and cultures of the institutions themselves must be conducive to such work in bridging esoteric forms of expertise, recognizing that it all takes time. Large organizations with functionally distinct and geographically scattered units face special challenges: How will the investigative work of different units be coordinated? How will the differences of training, orientation, and status of different actors be overcome without them working together over an extended period? Will institutional cues and incentives support investigators in doing good work or deflect them from it? The significance of such factors for the epistemic effectiveness of organizations is borne out by case studies of "blindness" to what should have been evident.[3] An epistemically well-ordered society requires epistemically well-ordered institutions, and the latter requires institutional norms and structures conducive to the efficient production of knowledge.

The second point is that a system of public knowledge geared toward addressing matters of public concern would also undertake investigations that are responsive to both the public interest and to scientists' judgments of where the most promising paths for future research lie, and it would frame the knowledge produced in ways suitable for guiding public and personal choices. We need institutions of public knowledge that can in some way transcend the boundaries between public interest and scientific expertise as well as the boundaries between diverse domains of expertise. This implies that a significant portion of sponsored research would be pointedly focused on matters of public concern, rather than simply flowing into reserve funds of knowledge that may prove at some time and in some way to be of value. Philip Kitcher has envisioned an idealized scenario by which the direction of scientific research would be determined neither by experts alone nor by popular demand:

Scientific significance accrues to those problems that would be singled out under a condition of *well-ordered science*: science is well-ordered when its specification of the problems to be pursued would be endorsed by an ideal conversation, embodying all human points of view, under conditions of mutual agreement. ... the ideal conversationalists are to have wide understanding of the various lines of research, what they might accomplish, how various findings would affect others, how those others adjust their starting preferences, and the conversationalists are dedicated to promoting the wishes others eventually form.[4]

Kitcher says that by this standard an actual society's pursuit of scientific inquiry would be well-ordered only if its research priorities are arrived at through deliberations modeled on this ideal and only if the priorities would be approved by ideal deliberators.[5] Those who have been pursuing research would tutor other participants in the deliberation so that they all know what possibilities of investigation are the most scientifically promising. Deliberators would then assess those options in light of their own concerns, and the conversation would ideally progress toward a common or most acceptable set of priorities. Kitcher allows that no actual conversation like this is possible, but how else are we to imagine the deliberations by which the production of knowledge in the public interest would be guided? We could perhaps imagine representatives behind a *legislative* veil of ignorance being privy to not only existing consensus science and the nature and extent of various obstacles to people living well, but also to expert judgments concerning the prospects for possible lines of research, as required by Kitcher.[6]

Despite the obstacles, there are in any case institutions that have already brought scientists together with representatives of public interests to mutually inform each other's endeavors.[7] Sometimes referred to as *boundary organizations*, one of their functions has been to "translate, negotiate, and communicate among the multiple parties on both sides of the science-use nexus."[8] A good example of such an organization is the Arizona Water Institute, which has brought together Arizona water managers and hundreds of university-based water researchers. It engages stakeholders in a variety of ways, including through intermediaries who can explain the science and through scenarios of possible water futures that allow managers to project the consequences of decisions.[9]

The third point is that it is important to the work of epistemic institutions that they employ people who possess the requisite epistemic virtues, capabilities, and understanding and that the conditions of work they provide align with the fulfillment of those attributes in *good work*. This brings us to the next form of basic institution to be addressed.

Educational Institutions

Just societies would provide all of their members with educational institutions, the basic function of which is to promote forms of development conducive to living well. They would do this by promoting the acquisition of understanding, virtues of intellect and character, and capabilities, in circumstances favorable to expressing these developing attributes in rewarding activity. Constitutional framers who considered the functions of institutions in the abstract while knowing little of human nature might imagine that the *acquisition* of good qualities and the rewarding *expression* of good qualities could be separated so as to occur in different institutional contexts. They might imagine that acquisition of good attributes could be motivationally sustained without the inherent rewards of expressing those attributes in action, but this is developmentally and motivationally all but impossible. Time spent in settings which are not favorable to engaging in inherently rewarding activity is also, all else being equal, a lost opportunity to engage in activities of a life well-lived. Therefore, to the extent that it is even possible to sustain high-quality learning, public service, or labor that is not inherently rewarding, such learning, service, and labor can only be collectively choice-worthy if they are so efficient in securing prerequisites for the activities of lives well-lived that the opportunities for everyone in the society to engage in such activities and live well are thereby improved.

Self-determination theory is the most comprehensive theory and body of research on motivation yet developed, and it has shown that the satisfaction of basic psychological needs for autonomy, competence, and relatedness is essential not only to personal well-being but to the effective internalization of goals and values.[10] Organismic integration theory (OIT), a key component of SDT, distinguishes four grades of internalization or adoption of motivating goals and values. These are markedly different in the qualities of actions they engender, independently of differences in "quantity" of motivation, and they range from the least to most autonomous: controlled, introjected, identified, and integrated.[11] Activity engaged in less autonomously is characterized by less energy and persistence, less pleasure, more internal conflict and stress, and a higher rate of errors. Action owing to *controlled* motivation is stimulated by an external force, such as a superior's direct orders, threat of punishment, or offer of a reward. Motivation is *introjected* when threats of punishment, shaming, or other external sanctions are internalized, and people act so as to avoid these internalized threats without accepting the value or goal as their own. Action arising from *identified* motivation is attributable to values or goals a person identifies with or accepts as her own, on the strength of reasons and a perception that she is free to accept the reasons and embrace the values or goals or not, as she sees fit.[12] SDT classifies this as a form of autonomous motivation. Further self-examination and self-regulatory striving would yield *integrated* motivation, in which the values a person identifies with form a more coherent whole and would be more seamlessly deployed in response to the complex particulars of situations.

The autonomy-supportive contextual factors identified as favorable to autonomous motivation include the offering of a *rationale* that is meaningful to the learner, respectful *acknowledgment* of the learner's "inclinations and right to choose," and a manner of offering the rationale and acknowledgment that *"minimizes pressure and conveys choice."*[13] By contrast, it has been established through nearly one hundred experimental studies that the use of rewards and penalties as motivators diminishes intrinsic motivation and the experience of autonomy, yielding controlled motivation that may be intense but less predictive of favorable engagement, quality of performance, and well-being.[14] SDT implies that for education to succeed it must be predicated on enabling students to satisfy all three basic psychological needs. It must foster a nurturing and cooperative community of learning, and it must structure learning and the learning environment in ways that allow students to develop and exercise their own judgment and meet attainable challenges that allow them to experience a rewarding growth of competence. Without such needs support, there is little prospect of students accepting a school's goals and values as their own or making the efforts essential to meaningful learning.

Educational institutions can best promote forms of development conducive to living well in a needs-supportive setting by coaching students in structured activities through which understanding, capabilities, virtues, and appreciation of the value of things develop, all in connection with forming interests and identities and finding meaning and direction in life. This has been referred to in philosophy of education as *initiating* learners into practices pertaining to things of value.[15] These things of value to which students may develop attachments include powerful ideas, artistry in diverse domains of performance, qualities of craftsmanship, communities of civic, professional, or athletic practice, aspects of the natural world, and so on. An introduction to these goods expands a student's understanding of value and opportunities for self-directed activity while offering resources and standards for critical thinking and judgment. Finding the activities of one's life meaningful is an essential aspect of living well, and such learning offers opportunities to live well in part by expanding the range of goods that might lend meaning to a life and doing so in conjunction with nurturing capabilities through which students can relate to those goods in significant and productive ways. Initiation into a diversity of valuable and rewarding practices provides opportunities for fulfilling intellectual, productive, and social potentials in admirable and satisfying ways.

Initiation into practices of inquiry, evaluation, and self-examination is essential to the cultivation of both intellectual and moral virtues, inasmuch as good judgment is a defining aspect of true moral virtue, and essential to competent self-determination in life, as in countless circumscribed spheres of activity. The formation of good judgment was long regarded as a central or overarching aim of a liberal or general education, but it is an admittedly daunting task.[16] A variety of considerations and bodies of research suggest that the cultivation of good judgment—of self-governance in accordance with good judgment—should begin in forms of character education that orient children to thinking things through before acting. Such cultivation would involve instruction in critical thinking and guided practice in analyzing historical and fictional case studies in judgment and choice. It would use integrated curricula and cross-curricular, inquiry-based learning that provide experience in bringing the resources of diverse disciplinary and analytical frameworks to bear on matters of importance to students' present and future lives.

It would ground all of this in school cultures shaped by what is known about the role of basic psychological needs satisfaction in sustaining student effort and the role of socially nurturing, autonomy-supportive, and optimally challenging school environments in meeting those needs.[17] Small schools can more easily function as cooperative learning communities that meet children's relational needs, and the strengthening of teacher-student relationships lends itself to educationally productive forms of character education and classroom management—ones that cultivate virtues of self-governance by engaging students in forms of self-reflection and self-management, rather than relying heavily on extrinsic rewards and punishments.[18] Moral virtue and the motivation it involves are acquired through experiencing a nurturing and just social environment and through *guided practice*.[19] The supervision and coaching of practice would call learners' attention to factors that are relevant to decisions, provide a related moral vocabulary and explanations, and guide them in exercising the forms of discernment, imagination, reasoning, and judgment on which good decisions are based.

In the interest of promoting development that is not only directly conducive to students' acquisition of attributes essential to living well but also instrumentally conducive to maintaining and strengthening a society's capacity to enable all its members to live well (as our eudaimonic principle requires), the practices into which students are initiated should include those needed to sustain the institutions essential to a just society. There should be variety that is favorable to diverse children finding what is conducive to their own flourishing, but also a favoring of practices that are important to a functional, sustainable, and legitimate social, civic, and global order conducive to everyone living well.

Workplaces

To the extent that the nature of workplaces is feasibly subject to public choice, societies have a collective interest in adopting policies to bring workplaces into alignment with employees doing *good work*. What is true of institutional contributions to the quality of school work is similarly true of institutional contributions to good work generally; institutional factors influence both the outward qualities of the work and how personally satisfying the work is for the person performing it.

Frederick Herzberg's international studies of work in the 1960s found that work is not satisfying unless it provides the worker with opportunities to use her own judgment and abilities and to experience accomplishment. Financial rewards and perks were found to mediate the extent of worker *dis*satisfaction, but rarely accounted for satisfaction.[20] These findings led Herzberg to suggest that financial compensation and other extrinsic motivators are the focus of excessive and counterproductive attention—a distraction from the factors that matter most to happiness at work and to the rest of our lives insofar as they are affected by our work. Research through the intervening decades has reinforced and provided more insight into these findings. Howard Gardner, Mihaly Csikszentmihalyi, and William Damon found in a major study of work undertaken in the 1990s that work satisfaction and burnout are strongly related to fulfillment and frustration of aspirations to do good work. Journalists in the United States suffered low morale because the conditions of their employment aligned poorly with their aspiration to present "objective facts ... thereby, empowering the public to make adaptive choices," whereas geneticists were much happier in their work because the terms of their employment generally allowed them to enjoy the inherent rewards of scientific inquiry without compromising their integrity.[21]

SDT and related work in eudaimonistic psychology go a long way toward explaining why alignment with doing work that satisfies ideals of excellence, integrity, and public service would be essential to being happy in one's work. A lack of alignment would frustrate the positive fulfillment of human potentials associated with the satisfaction of basic psychological needs. Not being allowed to exercise one's journalistic capabilities, judgment, and virtues in the course of informing people about matters important to their interest would frustrate the fulfillment of all three basic psychological needs in service to others. A large body of research has shown that on the whole people do not feel free at work, that they feel less free and more controlled when salary and other extrinsic rewards play a more dominant role, and that time stress associated with long and unpredictable work hours not only frustrates autonomy needs but undermines the quality of relationships and life satisfaction away from work. The more salient extrinsic rewards are, the more they *displace* intrinsic motivation to do good work, making employees less autonomous and happy in their work, more error prone, less energized, and less inclined to work above and beyond what is strictly required.[22] The more that employers force employees into exploitative or heartless relationships with others, the more the employees experience stress in the frustration of their relational needs, even if they have internalized competitive values. An implication of SDT is that it is not possible to fully internalize values that directly frustrate the satisfaction of any of the basic psychological needs, including the need for mutually affirming relationships.

There may be many aspects to work being admirable, associated with a variety of forms and measures of excellence, but we should particularly emphasize ethical integrity as a form of excellence and aspect of good work. The expectations and provisions of workplaces should align with the requirements of ethical integrity and social reciprocity formulated in the eudaimonic principle by offering opportunities to engage in activities of a good life that contribute to others having such opportunities. Good work contributes to a world of opportunity to live well, giving scope to workers' capabilities, understanding and judgment, and sociability. By this standard, compatibility with sustainability is a defining aspect of good work.

The extent to which workplaces are feasibly subject to public choice in the pursuit of these ideals is an open question, but if societies must collectively tolerate work that cannot be made inherently rewarding, then a lesson of this section is that work that is not inherently rewarding should be compensated not with outsized financial rewards but in kind, through compensatory opportunities to engage in meaningful and rewarding activities. One obvious way to achieve this aim would be to reduce work hours while in some way ensuring that income does not fall below a threshold adequate to living well—a strategy for improving opportunities to live well that may in any case be essential to a better allocation of work in a world that cannot afford to rely on economic expansion to provide adequate employment for those who need or want it.[23]

A Model of Sociopolitical Complexity and Collapse

Socioeconomic systems tend over time to exhibit greater *complexity* (role differentiation) and *stratification* (hierarchical inequality). The growth of such complexity entails rising energy requirements, as well as exponentially rising information and coordination costs associated with widening chasms of understanding between growing numbers of specialists with increasingly esoteric forms of expertise. It also entails an evolution of social roles, and therefore opportunities, toward ones that are increasingly specialized and only accessible through lengthening investment in superior preparation and credentials. The *structure of opportunity* is thereby altered, structure being a function of both the nature of the roles themselves and how they are related to one another with respect to stratification, terms of access, and other factors. As this structure is altered, it features ever-larger numbers of distinct social roles to occupy, but the arc of opportunity bends not toward equality, but away from it. Those who are at no disadvantage due to lacking a credential in one era may be greatly disadvantaged by lacking the same credential in an era in which it is more widely possessed and has become a common requirement of access. It may be impossible to generalize about whether the opportunity set becomes better or worse on the whole, but the complexity and costs of seizing the opportunities grow dramatically. The growth of associated socioeconomic stratification—and associated disparities of socioeconomic advantage and political influence—disposes sociopolitical systems to legitimacy crises, which are politically most commonly addressed through economic growth. In discussing sociopolitical sustainability in chapter 1, we noted that meaningful simplification in the interest of sustainability would almost certainly entail reversing the trend toward ever more specialized and stratified occupational roles and reversing the declining marginal return and rising costs—energy, information, compensation, and regulatory costs—associated with that trend, citing the work of Joseph Tainter, an influential pioneer of sustainability studies.[24]

In addressing these matters, we will begin with a summary overview of Tainter's explanatory model of sociopolitical growth and collapse and note some limitations of the model. We will then extend the model in ways that enable it to address some further important aspects of unsustainability pertaining to competitive social dynamics and how those dynamics are exhibited in the growth of educational systems mediated by market credentialism, the spiraling costs of adversarial social coordination, and failures to coordinate and focus inquiry and teaching on collective or public goods.[25] This will deepen what we will be able to say about the compatibility of epistemic and educational institutions with sustainability, the preservation of opportunity, and the nature of simplification in the interest of sustainability.

Dynamics of Growth

Investigations into the causes of systemic sociopolitical collapse have identified a variety of factors, ranging from deforestation, drought, and other environmental problems to invasions, catastrophes, and various forms of internal tensions and failures.[26] When we distinguished sociopolitical sustainability from other forms of sustainability in chapter 1, we defined *sociopolitical collapse*, following Tainter, as a "rapid, significant loss of an established level of socio-complexity," manifested in rapid loss of socioeconomic stratification, occupational specialization, social order and coordination, economic activity, investment in the cultural achievements of a civilization, circulation of information, and the territorial extent of sociopolitical integration.[27] The threats to sociopolitical sustainability we identified included ecological and throughput unsustainability, but also included social patterns that require unsustainable flows of material and energy. When such patterns are pervasive, there will inevitably come a time when those flows will not be adequate and decline or collapse becomes unavoidable, regardless of any damage the society's activities may cause to the natural systems and assets on which the society relies. In explaining the idea of a society being on a path toward overshooting a sustainable reliance on material throughput, we gave the example of pyramid schemes, which rely on flows of resources from a continually expanding base and which collapse when the expansion can no longer continue. We noted that when such schemes are systemic, as they seem to be in the increasingly integrated global civilization of the present age, the limitations of the planet's natural systems become the limiting factor—the base of the pyramid that cannot expand. The idea of a pyramid scheme is a helpful illustration of the concept of a structurally flawed sociopolitical system, but it is not the whole story.

Surveying a variety of promising explanations of societal collapse, Tainter notes that societies have often faced and dealt with environmental and resource problems, catastrophes, external conflict and invasions, and internal tensions and failures. Factors of these kinds may well play roles in the collapse of societies, but states and complex societies are problem-solving organizations, and their very complexity is a reflection of their success in managing problems.[28] To explain why societal collapses occur when they do, it is necessary to explain why the problems at hand defy the society's problem-solving capacities. Is the problem too big or too hard to diagnose? Is the society's problem-solving capacity adequate in principle but paralyzed by disagreement and failures to cooperate? The strength of the explanatory model Tainter developed in response to the limitations of prior theories is that it focuses on declining yields on problem-solving strategies over time, shedding light on why a society that has successfully managed its problems in the past may lose its capacity to do so at some point. In doing so, the model also explains why societies tend to become more complex over time.

According to Tainter, it is in the process of solving problems that sociopolitical systems tend to become more complex. There is efficiency in specialization when a division of tasks between distinct social roles and specialists who occupy those roles allows each to focus on fewer kinds of tasks and develop a higher level of expertise and proficiency. Will a society call upon its farmers to wage war as needed or maintain an army of highly trained professionals? If soldiering is not simply one of many tasks of people who are otherwise employed, will soldiers themselves be divided into distinct roles, such as archers, infantrymen, field commanders, and so on? If the problem at stake is how to consistently prevail in battle, then greater differentiation of roles allowing a higher level of skill with specialized equipment may solve the problem, provided the differentiated roles are well-coordinated. Societies, like armies, grow in complexity *as systems*, and the heart of complexity is the number of distinct social roles and accoutrements of specialization.[29] Additions of new roles entail the introduction of specializations that did not exist before, and new specializations come with new forms of training, skills, expertise, tools, practices, arenas of activity, and relationships with other kinds of specialists in a growing network or constellation of social types and practices.

With the growth of complexity also comes a growth of expressive capacity—artistic, architectural, literary, religious, political, scholarly, scientific, athletic, and so on—and this expressive capacity has historically been commonly deployed more or less in the expression of public purposes, a public voice, and signaling of sociopolitical legitimacy. This proliferation of social roles and growth of expressive capacity also provides more diverse opportunities for individual self-expression and advancement, and—although this is not part of Tainter's theory—it would be astounding if the proliferation of roles in complex societies were not driven in part by the initiative of individuals who find the prospect of new roles enticing. The flourishing and enduring manifestations of these forms of expression are known as *civilization*.

Costs and Hazards of Growing Complexity

If all goes well, a society will be cognizant of the costs associated with introducing new specializations, and the introduction of new classes of specialists will allow it to better address its needs and problems. The costs are substantial, and growing complexity also presents some hazards. There are many costs associated with a growing number of distinct specialist roles, including (a) *educational costs* associated with the acquisition of a growing number of progressively more specialized forms of expertise; (b) *compensation costs* associated with inducing people to acquire and make intensive use of the specialized expertise required for demanding new roles—intensive because it is easier to recoup the expense of lengthy training and specialist equipment by employing it more intensively; (c) an associated *increase in inequality* or socioeconomic stratification; (d) associated *problems of legitimacy*—that is, difficulty and cost in justifying the terms of social cooperation, motivating cooperation, and responding to internal tensions, resistance, and defection; (e) growing costs and difficulties of *efficiently coordinating* the actions of different kinds of social actors who may work and live in physically separated spheres and have increasingly specialized expertise, practices, and orientations; (f) *costs of transfers and movement* between the interdependent, interacting parts of a growing system; and (g) rising *energy demands* entailed by all of the above. A key tenet of Tainter's theory is that for all of these reasons, per capita costs and energy use rise dramatically with additions of complexity. This leads him to emphasize repeatedly that unless the requisite energy flows can be sustained, a decline from the established level of social complexity is bound to occur.

Societies generally strive to preserve the level of complexity they have achieved, using the strategies that have allowed them to increase energy flows and complexity. A second central insight in the theory is that even if problem-solving strategies yield growing returns early on, allowing a society to become more complex, investments in these strategies tend to suffer declining marginal returns, and these declining returns may leave the society unable to manage the problems it encounters. Faced with problems that defy its problem-solving capacities, the society may then collapse or experience a more gradual and protracted period of involuntary simplification. Estimating the collective or net social return on additions of complexity also can be difficult, making it hard to discern whether the returns on additions of complexity have declined to the point that they are negative, making the society as a whole worse off.

Sustaining energy throughput is a problem to solve much like any other in these respects, but one whose success is foundational to addressing all the others. Sustaining energy throughput over time tends to involve greater complexity, hence greater cost and investment of energy. This alone would make declining energy return on energy investment probable and a fundamental threat to sociopolitical sustainability, but, in addition to rising costs associated with greater complexity, the benefits deriving from additional complexity also tend to decline over time. The reason this can be generally expected is that if investment decisions are reasonably well-informed and rational, they will be sensitive to both cost and expected benefit. The best low-hanging fruit will be picked first, and efforts will proceed from there until no further pickings would be worth the trouble. The combination of rising costs and rationality in identifying and adopting the most cost-effective innovations first predictably yields a pattern of declining marginal return on investments in complexity over time, regardless of whether the additions of complexity are social, technological, or both—and the latter is often the case, because additions of technological complexity and new forms of expertise are often interdependent.

In chapter 1, we noted the history of declining energy yields for whaling and petroleum exploration and recovery, and Tainter documents patterns of declining marginal return on investment across a variety of sectors, from investments in the efficiency of engines to research on established topics and higher levels of education.[30] The later steps along established trajectories add refinements that are generally less significant in the larger scheme of things, however essential they are to taking the next step. The feats of invention that produce clever refinements of cell phones may be highly profitable, but the net social good they yield is surely negligible at best compared with the original invention and early development of the telephone and other inventions that are now merged with phones. To the extent that the investments in additional complexity are now subject to the logic of fashion and the frequent replacement of manufactured items this entails, the wider impact on prospects for human well-being could well be negative.

Societies may drift into continuing to invest in greater complexity even when the net benefit of doing so has passed beyond negligible and actually makes things worse—and it can be very hard to judge when that point has been passed. Americans, as Juliet Schor has argued through a series of powerful books, are born to buy, overspent, and overworked.[31] There are reasons to doubt that the benefits of high-speed, fashion- and status-conscious consumption are adequate compensation for the stress and the time poverty that Americans endure, but these circumstances of life are not favorable to judging the trade-offs.

Competition and Collapse

Tainter notes that societies may also be compelled to enter this hazardous territory of diminished and negative marginal return on accumulating complexity by *peer polity competition*, the paradigm of such a competition being an international arms race. Threatened with foreign domination, "complexity [at a level comparable to one's competitors] must be maintained regardless of cost," Tainter writes.[32] Escalating aggressive and defensive measures follow the characteristic patterns of problem-driven complexity and cost while often leaving the competitors worse off—poorer and less able to respond to other problems and maybe less secure against aggression (or able to profit through threatened or actual aggression) than before the race began.

In Tainter's recounting of the collapse of the Western Roman Empire, it sustained favorable energy flows for some time by overrunning weaker neighbors with bountiful treasuries (accumulated stores of energy, in effect). As it grew and had to bear the expense of governing and defending a larger territory, it more often and predictably faced opponents that were better prepared, had less wealth to plunder, or both. Once conquered and looted, its territories yielded only the energy flows that could be extracted through taxes as a share of ongoing farming—a shift from a mining to a farming energy economy, in effect—and it increasingly faced the problem of farmers abandoning their land as the imposition on their livelihood became unbearable. Unable to sustain energy flows adequate to coordinating and defending a complex, interdependent sociopolitical system spanning an immense geographic area, the empire collapsed. The lessons of this story, played out in subsequent publications, include the significance of transitions from high-gain "mining" energy economies to low-gain "farming" energy economies.[33]

Limitations and Extensions of the Model

The Collapse of Complex Societies is very much a study of ancient societies. Tainter has elaborated the theory's implications for energy transitions in the modern age—the transition from wood to coal in the nineteenth century and the terminal decline of fossil fuels in the twenty-first century—but other aspects of modern industrial societies could be fruitfully addressed as well.

The first and most important is that a primary focus on ancient societies invites the idea that all costly social coordination—all governance, broadly speaking—is managed by a state apparatus. Tainter dwells at length on the taxes and currency devaluations through which successive Roman emperors funded the rising costs of a state apparatus complex enough to manage a growing, far-flung empire, but the theory itself implies that, in one way or another, an increasingly complex society will bear rising per capita coordination costs, even if its elite do not indulge in endless luxuries. The addition of new social roles and all they entail will only be efficient or yield collective or net social benefit if the activities of those playing the roles are properly coordinated. Whichever system does the coordinating, it will cost something, and the per capita costs will rise as the society becomes more complex. In the modern industrial age, much of the coordination or governance is undertaken by business corporations, which face much the same fundamental challenges of governing their own complex moving parts as state governments face in coordinating a society. An obvious implication of the theory is that the larger and more complex corporate operations grow, the larger the per employee governance costs will be, and these costs will be borne by employees and the public through lower salaries, higher prices, or more costly externalities than those of a business that could somehow operate as well without any corporate management costs.[34]

A second limitation of the theory is that it does not consider the ways in which internal analogues of peer polity competition may emerge in modern societies. It would seem to be characteristic of the present age that governance within societies is internally divided and often adversarial. Having created modern business corporations and allowed them to grow without limit in global reach, capital accumulation, legal capacity, and social and political influence, the governments of contemporary societies are in many cases eclipsed by and substantially hostage to the priorities of multinational corporations. Tainter's theory would predict an increasingly costly internal public-private competition to control the patterns and terms of cooperation within the society. It would also predict that the side that falls behind in this informational and regulatory arms race will lose and will no longer determine the rules by which the society or some portion of it operates. The notion that markets are governed by a costless invisible hand invites the idea that efficiency requires governments to lose these battles, but the efficiency or effectiveness of markets in serving the public interest depends on rare conditions of optimal political governance. Limiting the size of business corporations so as to preserve a sufficient number of independent market players is essential to the survival of free markets, and an unacknowledged implication of Tainter's theory is that it would be a necessary—though not sufficient—step toward societies reclaiming control of their own fates.

Mass school systems may give rise to another internal analogue of peer polity competition that bears scrutiny. Tainter argues plausibly that the social return on investments in education decline at higher levels of education but does not bring his account of the dynamics of complexity and cost growth to bear on the competition-driven pursuit of further and higher credentials. To the extent that cost growth is a product of peer competition for coveted opportunities, his theory suggests it may come unhinged from progress in solving the society's problems and yield insignificant or negative net social benefit beyond some level of expenditure.

These unexplored implications of Tainter's theory present opportunities to extend it in potentially fruitful directions. We offer some modest extensions in the subsections that follow, beginning with the competitive pursuit of credentials and the competition-driven growth of the US system of education and others modeled on it, which we term *market credentialism*. This pertains to growing complexity and cost at the interface of school and work. We will then address some patterns and costs of *adversarial coordination*, or competition-driven cost growth at the interface of the public and private spheres, and will conclude with some observations concerning the net social return on research and education.

Market Credentialism

Vast and overshadowing private fortunes are among the greatest dangers to which the happiness of the people in a republic can be subjected ...

Now, surely, nothing but Universal Education can counter-work this tendency to the domination of capital and the servility of labor.

—Horace Mann[35]

Horace Mann, prominent leader of the common school movement that ultimately gave the United States its system of universal public secondary education, wrote these words in 1848, at a time when few Americans earned a high school diploma and few if any of those who did were poor. It was easy to imagine then that universal secondary education could eradicate poverty and the domination of one social class by another, but by the 1960s it had become evident that it was not the great equalizer many had supposed. What happened instead, as high school completion rates reached about 70 percent, was that a high school diploma became both an essential prerequisite for obtaining work, being so common that employers expected it, and ceased to confer much advantage upon those who had it, except as a ticket of admission to college. The high school earnings premium collapsed, and dropping out of school was more or less simultaneously identified for the first time as an important social problem.[36]

In these circumstances, the obvious reason that a high school diploma would not provide groups of "last entry" into the school system with the same advantage it provided others before them was the saturation of the labor market with high school graduates.[37] By the 1960s, schools, colleges, and universities with all their burgeoning graduate and professional schools had come to function as one *integrated hierarchical system*, outside of which there were few and distinctly inferior opportunities to acquire meaningful preparation and certification for work.[38] An aspect of it being one integrated hierarchical system is that having an advantage in securing positions higher in the socioeconomic hierarchy is largely a function of securing credentials at higher levels of the educational system and from higher-status institutions. Another aspect of it being such a system is that when the labor market is saturated with credentials at one level of the system, the only value a credential at that level will have is in securing admission to a program at a higher level. If it were not a hierarchically integrated system in this way, and high school diplomas retained socioeconomic exchange value outside the educational system, then the implications for equal opportunity would not be as dire, and Mann's vision might be substantially fulfilled.

The implications of *contest mobility* in an integrated hierarchical system did little to discourage theories that assumed educational systems must have a function that was "secondary" to the structure of economic relations and that their function was to replicate that structure—to replicate the patterns of domination at which Mann took aim.[39] Nor have such implications discouraged theories that share the same basic assumption but diagnose the problem as a failure of schools or students' home cultures to play their assigned roles in developing students' human capital adequately. There are two respects in which the basic assumption is misguided, the first of which is to think that the part of the US educational system that explains the problem is not itself an arena of economic relations. If there is great expense, collective futility, and frustration of equal opportunity in a system that becomes as complex as the US educational system has become in response to the competitive advantage-seeking of students, it is not because public elementary and secondary schools are failing in their assigned roles in this integrated hierarchical system: It is because the United States has a robust market in the provision of higher education and the assigned role of primary and secondary schools is to enable every child to go to college.

Designing primary and secondary schools to enable every child to go to college has the apparent merit of being essential to achieving fair equality of opportunity as Rawls defines it and equal opportunity (EO) as Americans by and large conceive it. How else could young people develop their talents so as to qualify for whatever positions or offices they desire, regardless of their social class origins?[40] Similarly, a robust market in higher education has all the evident merits of free enterprise, consumer choice, and outstanding accomplishment in research. Yet together these structural features of the US educational system yield competition-driven investments in complexity, the net social benefit of which is unclear. They make universal primary and secondary education not the great equalizer, but the vast receiving end of a funnel through which all must pass to succeed. Those who begin ahead will by and large remain ahead and, like the molecules of a fluid moving through a constricted space, they will be prodded along at ever-greater speed by those pressing in behind them until they exit the funnel's far, narrow end. Privileged students call this *the race to nowhere*—a race in which they are so busy staying ahead that they have no time to think about where they are going.

In a passage that brings together concerns about inequality, misdirection of student focus, and declining net return on investment, historian David Labaree wrote the following in 1997:

The inner logic of the credentials market is quite simple and rational: educational opportunities grow faster than social opportunities, the ability of a particular diploma to buy a good job declines, so the value of educational credentials becomes inflated. ...

Credential inflation affects schools by undermining the incentive of students to learn. ... It [also] promotes a futile scramble for higher-level credentials, which is very costly in terms of time and money and which produces little economic benefit. ... the credentials market continues to carry on in a manner that is individually rational and collectively irrational.[41]

Why do educational opportunities grow faster than social opportunities? Labaree cannot be referring to the system of primary and secondary education that had grown enough to accommodate every child in the United States by the 1960s. He can only be referring to institutions of higher education and their enterprising history of finding opportunities to expand and offer new and higher degrees. At baccalaureate and postgraduate levels of the US educational system, there is no established limit to the number of institutions competing for students and tuition revenues, and there is no limit to the number of students upon whom they may collectively confer degrees. Although there are many public institutions of higher learning as well, they are often similarly dependent on attracting students whose families can pay tuition charges. This is a market system in higher education, and in such a system the educational opportunities and availability of earned credentials will grow until demand plateaus, and the demand may far exceed the supply of corresponding *social opportunities*, to use Labaree's term. Individual institutions need not concern themselves with this excess so long as they can recruit enough students of their own, and they will prefer that the programs of competing institutions be the ones to contract or close their doors if the market for a credential contracts. As the supply of a credential exceeds the opportunities that can be secured with it, the "exchange value" of the credential falls—there is *credential inflation*—and students with the means and inclination will seek further, higher, and more differentiated degrees. The market value of undergraduate degrees falls as such degrees become prevalent, just as the market value of high school degrees fell as high school degrees became prevalent, and the cycle does not stop there.

Large differences in family resources and the freedom of educational providers and consumers to pursue their own interests will ensure that by and large the groups of last entry into the system will remain as they began, at the bottom of a stratified structure of opportunity. This pattern of growing cost and complexity in a stratified market system (an integrated hierarchical system) of educational providers is what we mean by *market credentialism*. It is a pattern of competition-driven cost growth not unlike an international arms race, and Labaree's references to futility, paucity of economic benefit, and collective irrationality accord with Tainter's theory of declining marginal return on complexity. Labaree's reference to credential inflation undermining students' "incentive" to learn is also consistent with our own assessment of the eudaimonic and epistemic deficiencies of the system. These concerns about equal opportunity, declining net return on investment, and misdirection of student focus all warrant comment.

Equal opportunity. We have not yet introduced the second mistake involved in assuming that educational systems must have a function secondary to the structure of economic relations; it is time to do so. To say that those born to lower social classes will remain as they began, at the bottom of a stratified structure of opportunity, is still to understate the effects of pursuing both equal educational opportunity and market freedoms in higher education in the way these have been pursued in the United States. The basic point is that an effect of stimulating a credentials arms race by embedding universal secondary education within an integrated hierarchical system has been to stimulate the creation of proliferating arrays of new specialists and forms of expertise that have profoundly altered the nature of work and the structure of opportunity as these specialists have established new professions and found ways to make themselves useful to employers. This is a leading conclusion of recent work on the immense social impact of educational expansion, and it is consistent with Tainter's observations about peer competition and the relationship between social complexity and stratification.[42] The thesis that education is merely a secondary institution shaped by economic relations wrongly assumes that the introduction of new forms of social complexity could not often originate in the higher reaches of educational institutions and be propagated from there. They can and do, thereby altering the nature of work and creating a more complex and stratified structure of opportunity. This implies that higher education does not simply offer more advantage to the already advantaged; it plays a systemic role in shifting the opportunities available toward ones that are increasingly unequal and dominated by the analytical thought work characteristic of academic culture.[43]

Whatever the advantages may be of a system in which the growth of exotic new breeds of academic specialization creates a more complex and stratified world of opportunity, the dynamics of competition-driven educational growth and the role of higher educational growth in creating a more stratified structure of opportunity suggest that public efforts to raise the proportion of the public that enrolls in higher education in order to equalize socioeconomic opportunity are deeply misguided. There may be good reasons to expand collegiate enrollments quite apart from equal opportunity, of course. Global economic competition or the desirability of a strong general education for life and citizenship may require it. Expanding access to college has indeed been a focus of US educational policy, but to the extent that this effort succeeds it will lengthen the educational funnel through which all must pass and create a structure of opportunity ever more skewed toward rewarding business applications of analytical prowess. It will contribute to the declining labor market value of all such degrees, much as making secondary education universal contributed to the collapse of the high school earnings premium, and the contributions to human capital formation may do little to promote an expansion of opportunities— for one reason, because the expertise that is created may be employed in reducing business's reliance on labor. If the goal is equal opportunity, then policies that aim to increase the fairness of access to a collectively beneficial number of places in degree programs would be more sensible. In addition to a new policy, we need a new way to conceptualize equal opportunity to secure satisfying employment and live well. We need a way to conceptualize and pursue this aim that is not self-defeating in its application and is not predicated on equality of prospects with respect to a common pool of occupations.

Declining net social return. A society in which educational system growth is driven by competitive pursuit of credentials, which become all the more essential as their market value declines and costs rise, will face an internal counterpart of an external arms race. We can assume, following Tainter, that basic forms of education have the greatest noncompetitive or net benefit and that the marginal net benefit of further education generally declines.[44] As growth and the associated progress of specialization continue, the duration and cost of educational preparation in increasingly exotic forms of expertise lengthens and grows. Net social return on marginal investment in education will generally decline through the higher reaches of an education, and less of the more specialized education obtained will generate any net social benefit through deployment in any occupation or social role, whatever the individual competitive advantages may be of continuing to higher levels of education. This will be true for at least three reasons: First, it requires both luck and creativity to make occupational use of higher learning. Second, the general capabilities, virtues, and understanding of the world provided by a good general education tend to remain relevant through the rapid changes entailed by innovation and growing complexity, whereas more specialized knowledge and skills are more likely to become obsolete. Third, employers in the business world will seek to convert portions of the specialized expertise obtained for personal competitive advantage into corporate competitive advantage that may yield no net social benefit. There are a variety of ways in which this may occur (some noted in chapter 2), including through advertising, public relations, litigation, lobbying, mergers, and pursuit of market dominance that are insensitive to the public interest.

The predictable pattern is that as educational expenditures rise, a growing proportion of those expenditures will be absorbed in a zero-sum competitive struggle.[45] This would be enough to cause a flattening of marginal return on investment in education, but not in itself a negative return. Yet it could well be negative—or worse than an expensive act of collective futility. Whether it is a somewhat positive or zero-sum game or a negative-sum game depends on the immediate and long-term impacts on opportunities to live well. For the long-term, educational system growth is no more sustainable than general economic growth as a basis for providing everyone opportunities to live well; both involve rising consumption that will damage ecosystems and diminish opportunities to live well in the future. In the short term, it is far from clear that the burdens of profitable complexity are fully compensated by improvements in quality of life. People who are voluntarily working less and simplifying their lives evidently believe they are not.[46] Meanwhile, the immediate impact for many students consists of lengthening and costly years of education, which provide for many neither the desired market success nor the benefits of a system that is more needs-supportive and inherently rewarding. Owing in part to the abysmal quality of information provided to secondary school students concerning their prospects for success, only one in five remain on the high-status career path they envisioned in high school when they are seven years beyond graduation, and the rest are saddled with onerous debt and limited opportunities.[47]

An integrated hierarchical system in which the job of schools is to send everyone on to college might do better in this regard, but it is a system that gives schools little incentive to counsel students to forgo wasting their money and time on pursuing a college degree and career that is probably beyond their reach. The ideal of equal opportunity invites schools to encourage whatever dreams students may have, leaving it to the world to dispel 80 percent of those dreams through the ensuing years of costly educational investments by students and their families. Enabling students to better assess their ambitions without robbing them of hope would require not just better information but also actively providing and encouraging consideration of alternatives that are attractive and more realistic. If this occurred in school before students reach "the age of reason" and are recognized as having moral and legal standing to choose an occupation for themselves, it could be seen as a violation of EO or FEO. This approach would also require schools that are very different from those in the United States and much more serious about preparing students to qualify for more than college admissions.

Misdirection of student focus. Labaree is concerned that the focus on credentials and contest mobility in the US educational system undermines real learning that would serve civic ends and economic efficiency. We share this concern. The capacity of schools and students to focus on the collective and personally formative benefits of education is seriously undermined by an escalating, instrumental competition for credentials and preferred employment. The instrumental framing of the value of schoolwork often associated with pursuit of credentials and regimes of testing and accountability is also demotivating for all students, as they focus not on the inherent rewards of learning but on securing credentials efficiently.[48] We have already spoken to the importance of sound motivational understanding in education and repeat here for emphasis that if students are to grow into lives well-lived, they must experience progress in living well in school. They must experience themselves as becoming good at things, as members of a just and nurturing school community, and as learning to direct aspects of their lives using their own developing judgment. We highlighted these intrinsically motivating aspects of sound development as basic to the way just educational institutions would function. They are not incompatible with high schools preparing students for certification in the qualifications for work, if students have learned what they need to in a supportive setting, but they are hard to preserve in the face of frequent high-stakes testing as it exists in the US system. Motivating children with high-stakes tests and the promise of glittering careers at the end of many years of outdoing others is not a promising approach, and this is especially true of students who are not born to wealth and have few opportunities outside of school. A eudaimonistic or well-being approach would lead students progressively and in motivationally engaging ways into realistic life opportunities so that they could experience a growth of competence and living well that takes them into adulthood.

Complexity and the Costs of Adversarial Coordination

Educational institutions are market players in a system of market credentialism, and they interact with each other and with the business enterprises that can offer their students gainful employment through the medium of credentials. The growth of those enterprises in cost and complexity—including the complexity of their products and services—will also exhibit a pattern of declining collective benefit to the society. The requirements of survival and profitability in a competitive market will engender innovation and marginal improvements in products and services sufficient to influence sales, but at growing social, ecological, and resource cost and with diminishing to negative demonstrable gain in quality of life. Investments in advertising, public relations, lobbying, legal capacity, and private interest science will predictably rise, at the expense of transparency and public oversight in the interest of ensuring that the enterprises' activities promote the public interest.[49] The competitive struggles among business enterprises and the informational and regulatory arms race between the public and private sectors are further spheres of competition, and the growth of complexity and cost driven by these spheres of competition is abetted by a system of market credentialism that offers ambitious and well-positioned individuals opportunities for elite credentials. The patterns in these spheres are even more suggestive than those we have just examined of a system driven by internal competitive dynamics to continue along a trajectory of growing complexity despite a negative marginal return on investment.

Judging where a society is along such a trajectory is made all the more difficult by the very complexity of the systems involved. These grow more complicated not because anyone perceives collective benefit in such growth, but perhaps because a few individuals see potential for private benefit. Before the 2008 financial collapse, did even the inventors of the forms of mortgage investment instruments at the heart of the collapse fully understand how they would function? Perhaps they did, but it is fair to say that in a market full of such complex financial inventions, there was widespread failure to grasp how vulnerable the system was—and not for lack of trying. The fantasy of market self-regulation contributed to the crisis, as it did to others before it, but government regulatory capacity would in any case face very long odds in a context of adversarial social coordination or governance that allows corporate assets and capacity to grow without limit and be directed more or less at will into frustrating public oversight. It only makes matters worse that in the United States the public bears much of the financial cost and risk of educating students but is routinely outbid by private sector employers in the competition to hire the most talented graduates. It thereby underwrites its own disadvantage in providing oversight of business activities sufficient to have any idea of whether the public interest is served, let alone protect that interest.

Complexity entails a proliferation of specialized forms of expertise that (1) pose ever-greater challenges to the understanding and judgment of nonexperts; (2) are deployed to the advantage of private interests, often with the goal of making it impossible for other market players and the public to form sound judgments of what is in their interest; and (3) are sometimes so opaque to one another that the status of systems requiring their coordination may be impossible to judge in time to prevent system failures, even under norms of cooperation. Creating an epistemically well-ordered society that does not overstrain the educable limits of human judgment is a tall order, but it is essential to a society that is just, sustainable, and coherently self-governing.

Impaired Faculties

Stanley Katz, a President Emeritus of the American Council of Learned Societies, observed in the *Chronicle of Higher Education* some years ago that university growth and increasing specialization have led to a neglect of matters of public interest that do not fit neatly within ever-narrower and more technical specializations and have made it increasingly difficult to "mount plausible curriculums for undergraduates and even for graduate students."[50] These unfortunate consequences of increasingly specialized disciplinary and subdisciplinary "silos" has in the meantime become a staple of the literature of sustainability, cited as an obstacle to university research and teaching pertaining to matters of sustainability.[51]

As research projects advance and disciplines grow and divide into subspecializations, communications between the parts of traditional disciplines may become more difficult, just as communication between disciplines does, so that it becomes both uncongenial to many and all too rare that scholars attend to matters that do not fall neatly within the borders of their own specializations. As the inherent challenges of undertaking the necessary border crossings increase, rising standards of scholarly rigor within specializations—the standards on which the careers of leading scholars are built—may also deter many from undertaking messy cross-disciplinary work that has little hope of meeting those standards. Publication will be less certain and the rewards so unclear that even institutional encouragement to take on big and messy topics may move few to action. The teaching by research faculty in ever-narrower specializations is also likely to favor their own interests, leading to fragmented curricula that may be immensely stimulating but make no pretense of educating students for life or citizenship in the world they will inherit. These too are costs of the growth of complexity, understood as a growing number of distinct specialist roles, and the costs are potentially very heavy indeed.

Bending the Arc of Opportunity toward Justice

Must the arc of opportunity bend toward injustice? Nothing in the analysis we have offered suggests that it must, that it cannot be turned back on itself in the interest of justice and sustainability by acts of public will. We close out this chapter with some modest and unoriginal proposals for reorienting our understanding of equal opportunity, distributed social governance, and professorial collective ownership of curricula and the social value of research.

Overcoming Monopoly Credentialism

We noted previously that an aspect of the US educational system functioning as one integrated hierarchical system is that when the labor market is saturated with credentials at one level of the system, the only value a credential at that level will have is in securing admission to a program at a higher level. If high school diplomas retained socioeconomic exchange value outside the educational system, then opportunity would not require safe passage through the one bottleneck toward which all students in the United States are funneled—namely, college. Mann's vision of a high school that would end the domination of capital and servility of labor through universal free secondary education might be substantially fulfilled if occupational organizations themselves played suitable roles in the governance of vocational education, certification, and their coordination with general education. A partnership of labor and employer associations in overseeing occupational training and certification could be far more conducive to creating distinct, meaningful pathways to employment and responsible adulthood. It would require the establishment of educational spaces that would not compete for higher status by embracing the liberal studies and research that are valued by academic culture. An aspect of educational opportunities organized in this way is that the system of colleges and universities could not monopolize occupationally meaningful credentialing in the way it now does in the United States.

This describes some basic features of the German educational system, in which nearly two-thirds of students pursue secondary vocational education through either a dual system of apprenticeships and high school education or through employment-based training. Historian Hal Hansen writes: "While Americans look to college to secure a future for their children in the absence of a meaningful secondary certification system, Germans established hundreds of relatively attractive, legally regulated, skilled professions—not mere jobs—and organized an effective, highly standardized system for preparing the young for them at the secondary level. It trains not only bakers, hairdressers, auto mechanics, and machinists, but bankers, accountants, information technologists, engineers, librarians, and archivists—occupations Americans associate with higher education."[52] Committees of the Federal Institute of Vocational Education (*Bundesinstitut für Berufsbildung*) establish the training standards for 350 distinct occupations, and these committees give equal representation to employer associations, labor unions, the federal government, and states.[53] Standards are set high and continuously updated along with occupational profiles, ensuring meaningful relationships to the occupations that students seek.

This "system of semipublic self-government" or "parapublic" governance, as Hansen calls it, is the historical culmination of a progressive transformation of apprenticeship and craft guilds.[54] Although students are guided along different paths in the differentiated secondary system, the decisions are not irrevocable, and "vocational qualification is so attractive that twenty-eight percent of Germans with an academic degree (*Abitur*) enter the vocational system."[55] In contrast to their US counterparts, German students learn much more and are provided with far more useful information about their prospects, they do not bear the enormous personal expense of taking years to realize that their aspirations are unrealistically inflated, and they are not forced to compete directly with everyone else in a single hierarchical system that focuses public resources and private rewards in its higher extremities. Their enculturation into rewarding adult roles begins years earlier, is less dominated by an amorphous and indulgent youth culture, and leaves fewer "lost in transition."[56] The German system is far less a winner-take-all system than the US system, and it has not suffered the credential inflation and system growth that the US system has suffered.

The existence of the German system demonstrates that the structure of the US educational system is not the only structure an educational system can have and that it is possible to escape the costly dynamics of market credentialism. It does not follow that the German system is ideal, or that one like it could be created in the United States anytime soon, but the existence of the German system helps clarify what is essential to meaningful structural reform.

The fundamental objection that advocates of the US system have is that the German system does not allow all children to compete for positions of every kind when they reach the threshold of self-determining adulthood—the end of high school, more or less. It violates equal opportunity, as Americans understand it, and Rawls's principle of FEO, as it is commonly interpreted, because it guides students along different educational and occupational paths well before the so-called age of reason. The reality of respecting children's future autonomy before they have mature judgment (have "reached the age of reason," as Rawls says) is that adults do shape children's opportunities to learn what they might be good at, their impressions of what they can realistically hope for, what they admire, who they want to be like, and where they feel they belong. Schools have this shaping effect, whether they embrace it or not, and the best they can realistically do—and ethically should do—is give students diverse starting points for developing interests, capabilities, and understanding of what they might do with their lives, and be attuned and responsive to their inclinations and talents.[57] If a system operating on the German model did this well, it might honor children's right to an open future, but the inevitable difficulties in discerning differences of native talent against a background of unequal and differentiated prior education would make this all but impossible without policies that sharply limit inequalities in the socioeconomic status of children's families.

There is also the important question of how much general and civic education all children need in order to be adequately equipped for life, citizenship, and positions of leadership. An *excellent* general education through high school might be enough, but probably not in a system in which many children spend much of their time in apprenticeships away from school. To ensure the adequacy of general education for all children and delay tracking them toward different occupational outcomes before the end of high school, one could envision a dominantly academic general education through high school and replicating the strengths of the German system of differentiated occupational credentialing at the two-year community college level. One could also envision options for free or low-cost continuing adult general education in a noncredentialing division while limiting four-year college capacity to the number and variety of seats it would be collectively rational to subsidize or fully fund.

If emulating the German system of occupational credentialing in the US community college system were feasible, it would come much closer to fulfilling Mann's vision of a universal education that could overcome "the domination of capital and the servility of labor." The fundamental point is that beyond the education everyone needs, escaping the pitfalls of market credentialism requires an effective system of *differentiated* occupational education and certification that does not force all students to outdo everyone else in climbing the rungs of a system that is, from an occupational perspective, a costly test of endurance and family resources, in which students bear inordinate risk. For all the merits of the US system of higher education, it is not the one best place to provide at students' expense what many would have been happier to receive free in high school or community college: occupational qualifications and meaningful, workplace-based learning that provides a direct and enculturating path to good work.

The possibility of a reformed educational system along these lines suggests that EO and FEO do not presuppose an integrated hierarchical system of education and could be more feasibly promoted without one. This is an important result, because it suggests a basis for reforms that could promote synchronic EO and FEO while restraining the system growth associated with market credentialism and preserving a more favorable structure of opportunity over the long-term. Although FEO may not be fundamentally self-defeating in its implementation, however, it still does not provide a basis for conceptualizing the long-term preservation of opportunity to live well because it is predicated on the existence of a common pool of occupations for which everyone can compete, and occupations will evolve over time (even without the stimulation of a growing educational system).

As a starting point for conceptualizing diachronic distributive justice, our own *eudaimonic principle* suggests the following:

Eudaimonic Distributive Principle: Positions and offices are to be distributed on the basis of bona fide qualifications, and everyone is to have prospects that are as good as everyone else's, regardless of social class origin, to find and do good work and live well.

Good work contributes directly to a life well-lived and fulfills basic psychological needs, which is essential to and predictive of personal well-being. It requires autonomous endorsement of the occupation and social role as one's own choice, but it does not require that all other existing occupations have been equally available, as FEO would. It is not predicated, as FEO is, on comparisons of prospects with respect to a common pool of occupations, and it is in that sense closer than FEO to being a principle that would capture the kind of forward-looking intergenerational justice that is the focus of concern with sustainability. A corresponding principle of opportunity preservation can now be stated:

Preservation of Opportunity Principle: The individuals of future generations are to have prospects of living well that are as good on average as those of individuals alive now, regardless of when they are born.

The interpretation of this principle should be guided by our previous explanations of the universal requirements of living well, the role of sociopolitically sustainable institutions that are just and conducive to sustainability, and the fundamental roles of ecological and throughput sustainability in preserving the natural bases of opportunity to live well.

Redistributing Governance

Adopting relevant aspects of the German model of occupational certification would end monopoly credentialism and redistribute the governance of work in a way that would be favorable to workplaces being more just or compatible with the eudaimonic principle. This is one example of what we mean by redistributing governance so as to be more conducive to the preservation of just institutions and sustainability. The general point is that a property-owning democracy characterized by far less concentration of wealth and influence would be one in which there might still be adversarial coordination or contests of governance, but individuals and the public would have the advantages of better bargaining position and a more level playing field. The society could more closely approximate the ideal of being epistemically well-ordered.

Reforms of corporate law and much stronger and more widely considered enforcement of antitrust law (so as to focus not only on the notional interests of "consumers") would be essential. Efforts to reduce and limit concentrations of ownership would also be essential, not so much to ensure the universal adequacy of material prerequisites for living well as to redistribute governing authority. This should include the authority to which employees are subject in their employment, which is presently often in oligopolistic markets that make exit onerous if not unthinkable. Full transactional transparency is a cornerstone of individual autonomy and the autonomy of a self-governing public, and approximating it is essential to the forms of distributed social and economic coordination that would be conducive to sustainability. A lesson of Tainter's model of complexity and collapse, like Elinor Ostrom's research on self-organized distributed governance (introduced in chapter 2), is that governance through far-flung, complex systems is not only costly but insensitive to local circumstances and prone to mistakes, gaming, and legitimacy crises. This is no less true of globe-spanning business management than of political governance, and in the interest of both justice and sustainability it would make sense to reexamine the perceived efficiencies of scale in the business world.

Cross-Disciplinary Sustainability Research and Education

Although it is rarely discussed, the faculties of most colleges and universities have a collective delegated responsibility to fulfill their institution's educational mission, making them not just instructors in their chosen specializations but *trustees* with an obligation to collectively oversee and ensure the suitability of the whole of what students learn.[58] What we urge is simply that college and university faculties embrace this responsibility and recognize that this obligation is not fulfilled simply by ensuring that curricula are current with respect to the many diverse specializations in which faculty do research and often prefer to teach. Faculty "must learn from their colleagues in other disciplines in order to liberally educate students by teaching the connections between bodies of knowledge and the integration of knowledge."[59] They must also embrace a greater instructional focus on the integration of knowledge that bears on big questions and the challenges of the age—problems with which students must contend, if only as citizens. The implications of this collective curricular responsibility for an interdisciplinary topic such as sustainability are clear: Faculties cannot ethically ignore interdisciplinary subjects of importance to the instructional program of their school, believing that they have obligations only to their department or field of specialization. Not ignoring these obligations requires the participation of instructional faculty in ongoing cross-disciplinary conversation, adaptation of their own teaching and redirection of a portion of their scholarship in support of teaching as required by fair terms of collegial cooperation in providing a sound academic program, and cooperation with administration in the design of programs and direction of hiring to provide a sound academic program.

There is reason to be optimistic that reorienting education in these ways might be transformative in yielding more collective benefit for our investments in education and wasting correspondingly less in a quest for credentials. There is also reason to be optimistic that reorienting a share of research around cross-disciplinary topics of public importance could raise the net social yield on research investments, which might otherwise fund diminishing progress on long-established lines of inquiry.

The professional schools of universities also deserve attention in this regard, given their relationships to professional organizations and the role they play in preparing students for careers. There are signs of movement within some of the professions to incorporate principles of sustainability within their norms and codes of professional practice and to call upon professional schools to do far more to educate their students in principles and methods of sustainable practice.[60] UNESCO meanwhile has called upon higher education to provide sustainability teaching and research.[61] Like other institutions, professional schools should operate in ways conducive to sustainability, and they would most obviously do so by providing their students with education in sustainability, including the ethics of sustainability, as an integral part of the professional education they provide. Like all education, professional education should promote development conducive to students' living well. As professional education, it should do this primarily by enabling graduates to do good work in a profession. Understanding the context in which one lives and works is essential to living well and doing good work, and sustainability is an important and pervasive aspect of that context. The ethics of sustainability would be an inescapable aspect of the integrity and compatibility with the public interest that are essential to good work.

Conclusion

We introduced Joseph Tainter's theory of complexity and collapse in this chapter in order to address sociopolitical sustainability and the survival of institutions that are just and conducive to sustainability. The theory offers the sobering lesson that growing social complexity requires growing per capita energy consumption but yields declining marginal collective benefit over time. It implies that voluntary simplification may be essential to sustainability and that it would involve reducing both consumption and the complexity of the social system itself. It would involve individuals being more broadly competent or multitalented, focused less in specialist careers and relying more on the self-provisioning, or do-it-yourself mindset, that was much more common a generation ago.[62] This may be discouraging to some forms of ambition, but there is no reason to think that a less hurried life of being good at more things would be a less happy or less admirable life. We have also seen ways in which the conduciveness of institutions to the long-term preservation of opportunity may be entangled with and badly served by the way equal opportunity is currently pursued in the United States—that is, through economic and educational growth. The hopeful lesson we have drawn is that the pursuit of equal opportunity in the present need not diminish opportunity in the future. Indeed, it can facilitate it.

When we introduced Tainter's theory, we wrote that to explain why societal collapses occur when they do it is necessary to explain why the problems at hand defy the society's problem-solving capacities. We noted that a virtue of his theory is that it focuses on the returns on problem-solving strategies over time, shedding light on why a society that has successfully managed its problems in the past may lose its capacity to do so as time goes on. We asked in passing whether societies may face problems that are too big or too hard to diagnose and remediate, and we asked whether a society's problem-solving capacity may be adequate in principle but paralyzed by disagreement and failures to cooperate. These questions are at the heart of discussions of "wicked" problems in the sphere of public administration, and "wicked" is what problems of sustainability are often said to be. We will consider what the idea of wickedness might contribute to the understanding and pursuit of sustainability in the next chapter and offer a trio of case studies in energy, water, and food systems. Our purpose in presenting these cases will not be to solve the problems at stake in them but to illustrate the challenges of managing complex systems in more sustainable ways.

5 Managing Complexity: Three Case Studies

We have argued that there are reasons of both prudence and morality for broad cooperation to address climate change and other problems of sustainability, but in chapter 4 we glimpsed how difficult and costly such coordination may become with the growth of social and technological complexity. Another problem to contend with is that people who agree about the importance of sustainability may nevertheless differ in their judgments of better and worse allocations of the burdens of cooperation. Even if a uniquely fair principle of cooperation can be identified in the abstract, decisions must often be made in nonideal circumstances by decision makers who have limited powers and must contend with uncertainties and a public divided by divergent and competing interests, beliefs, and commitments. These decision makers must somehow bring together in one judgment the analytical resources of an open-ended array of sciences and the legitimate interests of stakeholders, addressing problems that may span multiple legal and political jurisdictions. Political theorists discuss some aspects of such decision making under the heading of *nonideal theory* or *nonideal justice*, but the term of choice in public administration, forestry and water management, and related fields is *wicked problem.*[1] The problems characterized as "wicked" are matters of public concern, typically involving interactions of complex social, economic, and environmental systems that are hard to predict and hard to control. It can be difficult to diagnose how these interacting systems will behave, in part because the social and economic systems are networks of stakeholders whose cooperation may be essential to solving the problem. The effectiveness and legitimacy of a plan of action may require the input of these stakeholders, and their ongoing cooperation may depend on a variety of factors that may themselves be altered by systemic processes, whether or not the stakeholders have a voice in determining a collective course of action. Problems of sustainability are often characterized as wicked, and sustainability itself has been characterized as a wicked problem that "cannot be solved, but only managed."[2]

Does the concept of wickedness shed light on problems of sustainability? What exactly is this concept? We will examine it in the context of the previous chapters and suggest a less metaphorical way of summing up the challenges of sustainability and how they might be overcome. We will then discuss three case studies in sustainability, pertaining to energy, water, and food systems. These cases will illustrate the challenges and offer some useful lessons.

Are Problems of (Un)sustainability Wicked?

The idea of a wicked problem was introduced in 1973 with reference to the failures of top-down "engineering" approaches to urban and social planning and has since been widely discussed in the spheres of public administration, applied economics, business, forestry and water management, and elsewhere, often with reference to complex systems.[3] Failures of top-down urban planning often reflected failures to understand and value urban communities as living systems, and as often as not involved bulldozing culturally vibrant expanses of mixed-use neighborhoods to make way for monuments of civil engineering like sports arenas and beltways that served nonresidents much more than those who were displaced. As products of top-down planning, these projects did not reflect the priorities of the communities most affected or consensus about the nature of a problem to be solved. In their seminal paper "Dilemmas in a General Theory of Planning," Horst Rittel and Melvin Webber distinguished *wicked* problems from *tame* problems and suggested that major public policy problems are mostly wicked.[4] Tame problems may be technically complex and far from easy to solve, but their solutions are not hostage to the dynamic complexities of living systems, there is no disagreement about the nature of the problem or what constitutes success, and they can be largely addressed by experts with little or no involvement of other stakeholders. In contrast, wicked problems are "dynamically complex, ill-structured, public problems" arising from complexity in both biophysical and socioeconomic systems.[5] Wicked problems are said to elude resolution because they are multicausal and intertwined with other problems. Attempts to tackle them result in unintended consequences, and they may change and manifest themselves in different ways in response to interventions or for other reasons, making them moving targets. They rarely fall within the responsibility or competence of any single individual or organization, making collaboration essential to managing them.

In his influential 1974 essay "Intellectualizing about the Moon-Ghetto Metaphor: A Study of the Current Malaise of Rational Analysis of Social Problems," Yale economist Richard Nelson posed this question: "If we can land on the moon, why can't we solve the problems of the ghetto?"[6] Landing on the moon was a straightforward challenge that was met using the tools of science and technology. Although an amazing technical achievement, the goal was well-defined and the criteria for success obvious: Astronauts would go to the moon, land on it, and return safely to Earth. Although there were several lunar landings and return trips, doing it once would have been enough. By contrast, the "problems of the ghetto" are ill-defined, multicausal, and pertain to the functioning of complex social, economic, and institutional systems. To solve such problems would be to make enduring changes in how those systems function, and the interventions available to governments are limited and deeply entangled with those very systems. Thaddeus Miller observes: "In order to understand how science and technology might contribute to sustainability, scientists, engineers, practitioners and decision makers would do well to consider whether a problem is more like the moon or the ghetto."[7] The answer favored by countless observers is that achieving sustainability is a problem of the latter variety, one that requires enduring changes in how human systems and institutions function.

A great deal has been written about how to recognize a wicked problem. Rittel and Webber suggested ten basic characteristics, which others have distilled into core dimensions that more simply distinguish tame from wicked problems. Economist Sandra Batie suggests two dimensions—level of uncertainty and degree of values conflict—with wicked problems scoring higher on both attributes.[8] Brian Head adds high levels of complexity as a third defining attribute.[9] Our own assessment of these efforts is that wicked problems may be better characterized in terms of *dynamic* and *decisional* complexity (figure 5.1). Both of these forms of complexity tend to be involved when the problem is an aspect of the functioning of human systems or the interface of human and natural systems. Poverty is a common aspect of socioeconomic systems and widely perceived as a problem, and we discussed in chapter 4 how a failure to grasp the dynamics of educational systems and systemic relationships between education and work encourages a misguided view of how poverty could be eliminated. Although the creation of a universal secondary school system in the United States was not without benefits, its role in an integrated hierarchical system created a

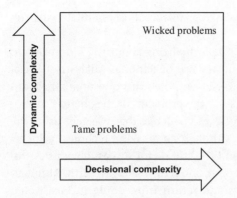

Figure 5.1
A two-dimensional model of wickedness

moving target and little social mobility. The history of attempts to eradicate unwanted parts of ecosystems, identified as pests, while retaining all the other parts reflects a similar unawareness of the dynamic complexities that are characteristic of ecosystems. John Sterman describes *dynamic complexity* as "the often counterintuitive behavior of complex systems that arises from the interactions of the agents over time."[10] We intend a wider meaning for this term that also includes dynamic interactions between human and nonhuman systems.

Joseph Tainter and Tadeusz Patzek discuss another aspect of complexity arising from "a proliferation of structure (more parts, more kinds of parts, more kinds of things to do) and the organization that is needed to make the parts and activities work together. ... The organization to make these parts work together (which they do not always do) comes in the form of social norms, beliefs, rules, requirements, regulations, laws, instruction manuals, and so forth, all the things that make people behave in a predictable way."[11] These "things that make people behave in a predictable way" are aspects of distributed governance by which people determine what to do more or less voluntarily, and a key aspect of wicked problems is that they cannot be solved by traditional command and control approaches in which a single competent manager or acknowledged authority can identify the problem and dictate clear-cut steps to solve it.[12] Attempts to manage such problems must cope with *decisional complexity* arising from the necessary involvement of diverse actors. These actors may bring divergent values, needs, and priorities and different types of knowledge, capabilities, and analytic tools to their roles, and they often must make decisions in the face of uncertainty about the conduct and states of mind of other actors.

Problems of sustainability are said to be wicked, and climate destabilization in particular is said to be a "wicked problem par excellence," a "super
wicked problem," and a "diabolical" policy problem.[13] Yet there is an obvious sense in which climate destabilization is a simple problem, despite its
global scale. Human beings are destabilizing the climate and acidifying the
oceans by burning fossil fuels, raising livestock, and clearing forests, and
the damage could be limited by *not doing* these things. We know well
enough how Earth's complex climate system works and what the effects of
our actions are. Far from wanting to alter the way the climate system functions so as to eliminate a perceived defect—as might be the case if we were
declaring a war on tornadoes, akin to a war on poverty—our problem is
how to avoid forcing it out of the favorable equilibrium that has made
human civilization possible. What makes climate destabilization not a simple problem to solve is that the activities causing it are systemic aspects of
human socioeconomic systems. The difficulties lie in the resistance of these
systems to rapid change of the kinds needed, beginning with the obstacles
to replacing one energy system with another noted in chapters 1 and 2 and
including the political and social challenges of negotiating and instituting
a global climate regime.[14] There is decisional complexity in establishing
the terms of compliance and dynamic complexity in bringing social and
economic systems into compliance.

If having these features qualifies climate change as a wicked problem,
then climate change is wicked—but is this an analytically useful category?
The literature of wickedness traces a salutary reorientation of approaches to
social planning and public administration, but a less metaphorical term for
the kinds of problems at stake would be *systemic action problems*.[15] If our
sampling of attempts to define wickedness is indicative of the current state
of thinking on the matter and our focus on dynamic and decisional complexity is accurate, then wicked problems would seem to be a class of *collective action problems* (see chapter 2). As such, their solution or management
requires steps to coordinate the actions of the various actors involved, and
Elinor Ostrom's work gave us a point of departure for understanding the
forms and prerequisites of such coordination. As we discussed, cooperation
on the basis of fair coordinative principles is more likely when there is accurate understanding of the systems, good communication, shared norms,
trust, competent leadership, and autonomy in the collective decision making needed to tailor implementation to local circumstances. Collective
action problems is a wide category, covering diverse conditions in which
individuals would experience better outcomes through cooperation than
through the most rational choices they could make in the absence of norms

of cooperation. This can be as simple as the difference between collectively deciding and not deciding to drive on a designated side of the road. We define the narrower category of systemic action problems as signifying problems in which the nature or functioning of systems must be altered and the effects of cooperative interventions would propagate through systems in ways that exhibit dynamic complexity. The dynamic complexity of systems often makes the diagnosis of such problems harder, and the need to alter the nature or function of the systems for extended periods challenges a society's capabilities to intervene more than causally and temporally bounded problems would.

Systemic action problems test human understanding, capabilities, and virtues of cooperation, in short. As we intimated at the close of chapter 4, they may be too hard to diagnose (understand), too big to tackle (with our limited capabilities), and too divisive to inspire an effectively coordinated response.

Coping with Systemic Action Problems

Understanding, capabilities, and virtues are the strengths that come together in the successful acts of individuals, and it is a merit of the concept of a systemic action problem that it should draw our attention to corresponding excellences of collective action. These collective strengths must be nurtured and orchestrated by a political art and a system of just institutions that rely on distributions of governance sensitive to both the requirements of regional and global coordination and the limitations of centralized and top-down management. There is no denying that humanity may encounter problems that defy its collective epistemic, creative, and cooperative capacity, but it would be premature to insist that problems of sustainability are already a lost cause. We addressed the nature of just epistemic, creative (workplace), and educational institutions conducive to sustainability in chapter 4, and we will address education at greater length in chapter 6.

We concur with Batie's observation that the production of knowledge by mainstream science has historically involved an inefficient one-way transfer of scientific knowledge into "a reservoir" from which society can draw as it sees fit.[16] Although there is undeniable merit in scientists having the freedom to pursue lines of fundamental research at will and leave the communication of science to others, public funding of science in the midst of present and projected suffering is properly sensitive to both the judgments of scientists and the public's priorities, and the flow of valuable knowledge into public policy could be better. Batie rightly points to innovative collaborations at the interface of science and resource management as one solution. We have also noted a variety of impediments to the public uptake of important research findings, including mistrust, misunderstanding of what is and is not sound science, manufacturing of doubt, and political intransigence. These are obstacles to the common understandings on which cooperation and widely distributed governance would be based and further deficiencies of public knowledge that might be remediated in the interest of sustainability.

In her introduction to the Australian Public Service Commission's 2007 report titled *Tackling Wicked Problems: A Public Policy Perspective*, Lynelle Briggs cited the growing complexity of problems confronting public managers, noting that tackling such problems is "an evolving art" and that "wicked problems require thinking that is capable of grasping the big picture, including the interrelationships among the full range of causal factors underlying them. They often require broader, more collaborative and innovative approaches."[17] We concur and would add that these collaborative approaches must accommodate uncertainty and be *adaptive*—or continuously evaluated and refined in response to what is learned through trials with provisional and partial solutions: Coping with problems of sustainability requires ongoing management that is collaborative, provisional, and adaptive.[18]

The problems of sustainability with which humanity must cope begin with climate disruption and include the dilemmas of growing and unsustainable demands on three fundamental requirements of modern life entangled with climate: energy, water, and food. We offer case studies concerning each of these vital provisions, and the cases illustrate many of the ideas introduced here and in the preceding chapters. Understanding the costs, hazards, and challenges of complex systems is important to managing human affairs in ways more conducive to environmental sustainability, and our hope in providing these illustrative cases is to lend greater clarity and vividness to ideas we have presented rather abstractly. The widely discussed concept of a *water-food-energy nexus* denotes the interconnectedness of these three vital provisions. Although sometimes criticized as an ill-formed concept, the word *nexus* signifies a coming together or connection of causal factors, drawing attention to the "big picture" causal interrelationships to which integrative resource management must be sensitive.[19] Decisions about energy systems often have implications for water and food systems, decisions about water often have implications for food and energy production, and so on. We will consider the 2010 Gulf of Mexico oil spill, drought and water management in Australia's Murray-

Darling Basin, and changing agricultural practices in the Mekong Region of Southeast Asia.

The 2010 Gulf of Mexico Oil Spill: A Cautionary Tale

Complex systems almost always fail in complex ways.
—Columbia Accident Investigation Board[20]

Systemic reliance on fossil fuels presents humanity with at least two dilemmas. Modern economies depend on them, but to avoid catastrophic climate change we need to leave the hydrocarbons from which they are derived in the ground. Yet even as the economics of energy favor continued reliance on fossil fuels and using cutting-edge technology to extend our reach to ever more remote environments, the energy yields on producing fuels from fossil hydrocarbons are plummeting. Investing the same energy and engineering resources in renewable energy systems would be a better long-term investment, but for the short-term economics of competing energy sources. Fossil fuels are thus both the lifeblood and the scourge of modern civilization.[21]

The illustrative episode on which we shall focus is the 2010 Gulf of Mexico oil spill, an accident involving cutting-edge technology that was considered accident-proof. What happened at the Macondo well reveals obstacles to accurate risk assessment in the management of complex technologies by socially complex teams of specialists operating within a complex top-down system of corporate governance in conditions of weak public oversight. It is also a vivid illustration of the potentially devastating environmental, economic, and social costs that may be incurred in the quest for previously inaccessible fossil fuels.

Why was BP, formerly known as British Petroleum, drilling in five thousand feet of water where only remotely operated vehicles can be sent to make repairs to reach a formation under immense pressure eighteen thousand feet below the sea floor? The answer is that the easily extracted deposits of petroleum have already been exploited, leaving only deposits that are in increasingly inhospitable and geographically remote regions. As the remaining reserves of fossil fuels become increasingly difficult to reach, extraction requires systems of increasing complexity. The growing energy investments entailed by this complexity ensure a declining energy return on energy invested, which is a measure of the net energy available after accounting for the energy used to extract and deliver a unit of energy.[22] During the early days of the oil and gas industry, readily accessible reservoirs on land yielded an EROEI of around 100:1.[23] By contrast, the EROEI of oil produced by drilling in ultra-deep waters has been estimated at less than 10:1.[24]

The increasing complexity of extraction technologies and varieties of associated expertise can also make it hard to assess and manage risk. Individuals and businesses with responsibility for different aspects of the Macondo operation made decisions on the basis of faulty assumptions about instrument readings and what others had succeeded in doing, under time pressure arising from the immense cost of deploying a system of such technological and human complexity. Tainter and Patzek characterize the resulting disaster as a prime example of a *normal accident*: "Normal accidents appear as if they are Black Swans, something that cannot happen. In fact, the very nature of complex technologies makes accidents probable. They are a normal byproduct of the operation of systems whose complexity is beyond human understanding."[25] The Macondo blowout illustrates not only the hazards of continued reliance on fossil fuels but also the limitations of remote command and control management, the challenges of coordinating the expertise and actions of players in a complex system, and the hazards of weak government oversight.

The Blowout and Its Aftermath

Just before 10:00 p.m. on the night of April 20, 2010, "a thundering explosion rocked the rig, the beginning of a terrifying night for the men who would survive one of the most harrowing disasters in the history of the oil business." Rig workers described a second explosion, even louder than the first, as "a tornado of fire, a nuclear bomb, a jet engine exploding … 'We all were sure we were going to die,' said one."[26] Drilling for oil in deep water is never easy, but no one was prepared for the explosions and fire that ripped through Deepwater Horizon (DWH), a mobile rig floating about forty miles off the shore of Louisiana.[27] The explosion killed eleven people, injured seventeen, and caused the worst accidental oil spill in history.

DHW was a mobile thirty-three-thousand-ton semisubmersible drilling platform that spanned the length of a football field and used a computer-controlled dynamic positioning system to keep it in place while drilling. On the morning of April 20, workers aboard DWH were relieved to be wrapping up their work on an exploratory well dubbed *Macondo*. Its target was hydrocarbon-rich rocks located twenty-three thousand feet beneath the water's surface, and the crew had experienced a number of problems, including loss of tools in the well and "kicks" (explosive intrusions of natural gas), prompting a BP engineer to deem it a "nightmare well which has everyone all over the place."[28] By then, the Macondo well was nearly six weeks behind schedule and $58 million over budget, and the crew was focused on finishing up and moving on to the next job. Drilling had been completed several days earlier, and the well had been lined with steel and plugged with cement. One final step remained: to conduct a series of tests to verify that the casing and cement would prevent the leaking of gas. At 10:43 that morning, a BP engineer sent an email describing a change to a crucial safety test, leading to confusion and debate among the workers. Less than twelve hours later, at 9:45 pm, the rig exploded in flames so hot that some of its steel parts began dripping into the water. A bubble of methane had escaped from the well, breaching several barriers to shoot up to the rig and explode, giving those who survived less than five minutes to escape.[29]

After burning for two days, DWH sank, breaking the pipe that connected the well on the sea floor to the floating drilling platform and releasing a torrent of oil. The spill and its consequences were vividly portrayed on television and the Internet through haunting images of gushing oil, petroleum-soaked pelicans, and tar balls on what had been pristine beaches as ten different techniques were tried and all failed to cap the well. On July 15, eighty-seven days after the blowout, the well was capped, and on September 19 it was permanently sealed.[30] There was considerable controversy regarding the magnitude of the spill, a debate fueled by streaming video from a camera on the ocean floor. The federal government initially underestimated the flow, creating an impression of either incompetence or lack of transparency about the scale of the disaster. Ultimately, estimates put the amount of oil entering the Gulf at about 4.9 million barrels (approximately 205.8 million gallons).[31] BP took responsibility for the initial clean-up efforts, with the US government joining in. Efforts to clean up the oil and mitigate its effects included controlled burns of surface oil, the use of chemical dispersants, and construction of sand berms. These were essentially the same methods used in response to the Exxon Valdez spill in 1989, suggesting that three decades of impressive technological advances in drilling for oil in environmentally sensitive regions had not been matched by any advances in modes of response to spills.

Innumerable scientific and legal investigations have probed the consequences of the spill, which can only be understood in the context of the Gulf of Mexico's uniqueness as a marine habitat and basis of economic and cultural life.[32] The Gulf is vitally important to the people who live on its coasts and derive their income from the fishing, tourism, shipping, and oil and gas industries that mingle in its waters. Its wider economic significance for the United States is evident in the facts that 90 percent of its offshore drilling takes place there and that its waters supply a significant portion of the country's annual seafood harvest. Moreover, the United States is not the only country to which the Gulf is important.[33]

Although President Obama called the spill the "worst environmental disaster in history," it also had a substantial social and economic impact. The oil ruined beaches, hurting tourism; decimated seafood catches and revenues; and killed the mangrove trees that provided nesting and breeding habitats and held together many Louisiana barrier islands that have now washed away.[34] Even before the spill, the Gulf of Mexico and many of the communities that rely upon it were under threat from coastal erosion and subsidence, sea level rise, hurricanes, invasive species, and a massive deoxygenated "dead zone" caused by phosphorus-rich agricultural runoff carried from farms in the Midwest to the Gulf by the Mississippi River—an example of a water system mediating a regional breach of a crucial planetary boundary.[35] Fully assessing the damage caused by the Macondo spill is a work in progress, and five years later scientists are finding both ongoing damage and resilience in the ecosystems of the Gulf.[36]

Risk Assessment and Decision-Making in Complex Systems
The challenges of drilling in deep water have been likened to standing atop a 110-story skyscraper and extending a long straw into a Coke bottle placed on the street more than 1,400 feet below.[37] Prior to the Macondo disaster, deep-water drilling was nevertheless considered "so safe and technologically advanced that the risks of a major disaster were infinitesimal."[38] The risks associated with complex systems have evidently been underestimated. The Macondo blowout resulted neither from mechanical failure as such nor from simple human error. It is instead best understood as system failure—a failure of collective action or governance in the broad sense in which we have been using the term. As we indicated earlier, the specific failures in this case pertain to the character and decisions of BP and its partner corporations, the complexities of drilling in deep water, and lax federal oversight.[39] These are failures of judgment in circumstances that overtax human judgment, and deficiencies of information flow and synthesis in a command-and-control system that is not built for epistemic efficiency.

The 2010 Deepwater Horizon disaster was only the latest in a series of accidents for which BP was responsible.[40] Noteworthy among the previous accidents was an explosion and fire at BP's Texas City refinery on March 23, 2005, in which fifteen people were killed and more than 170 were injured.

This refinery accident has been attributed convincingly to the "structural blindness" of a far-flung hierarchical organization in which cost-saving decisions to defer safety upgrades, made in luxurious comfort on one continent, were not responsive to workers' daily experience of the decrepitude and hazards of a seventy-year-old refinery on another continent.[41] Workers interviewed two months before the accident for an independent safety study reported an "exceptional degree of fear" of a lethal accident occurring, but no one with decisional autonomy gave the workers, refinery, or city priority over drawing as much wealth as possible to the corporate center.[42] BP's headquarters in London was separated from day-to-day operations at the Texas City refinery not only by seven layers of management, 4,800 miles, and a six-hour time difference, but by consequential differences of culture, education, social position, authority, and skill sets, all of which posed obstacles to the flow of knowledge through the organization and the adequacy of risk assessment and management.[43] It was from BP headquarters that chief executive John Brown ordered aggressive cost-cutting measures and from there that John Manzoni, chief executive of refining and marketing, made sure they were implemented, both men blind to the risks well-known to employees at the Texas City refinery. Those who grasped the risks firsthand had no voice, had no long-term commitment to the refinery or its crew, or were ignored in the interest of cutting costs. The managers in the best position to improve safety evidently conveyed a misleading impression of having everything under control and were promoted to locations closer to the charmed center of BP's global empire.[44] The Texas City accident is a lesson in the perils of remote management or governance that may value a place and its people only for their benefit to the whole— the kind of management tolerated by oligopolistic global markets in which public oversight is ineffectual.

The Deepwater Horizon accident was in many ways different from the Texas City refinery accident, but it can also be attributed to epistemic failures associated with complexity—the complexity not only of BP's system of management, but of the Macondo drilling operation and the diagnostic challenges it posed to a fragmented array of specialists with different forms of expertise, responsibilities, and employers. Drilling in deep water is a formidable task that requires the contributions of multiple actors drawn from a number of organizations, each contributing a different type of technical expertise.[45] BP purchased the rights to drill at the Macondo site, thus becoming the legal operator for activities undertaken there. BP's shore-based engineers designed the well and specified how it was to be drilled, but the company relied on a number of contractors to perform the actual work of drilling. BP leased Deepwater Horizon from another company, Transoceanic, at a daily rate of $500,000, and on the day of the blowout most of workers on the rig were Transoceanic employees.[46] A third company, Halliburton, was responsible for designing and pumping cement for the well. The Oil Spill Commission determined that a number of mistakes and errors in judgment by the three key companies played pivotal roles in the disaster. Investigators identified a "complex web of human errors, engineering misjudgments, missed opportunities and outright mistakes," including failure to assess and manage risk associated with several changes in well design, failure to replace a cement slurry in response to repeated tests that revealed problems with its design, failure to properly interpret a critical test that clearly showed that the cement had failed to seal the well, and failure to recognize and respond to early warning signs that gas has entered the well. Although numerous technical failures contributed to the disaster, each could be traced back to overarching failures of management.[47]

The commission also determined that ineffective government oversight had played a role. The promotion of industry self-regulation by the administration of George W. Bush was only marginally reversed by that of Barack Obama, and poorly funded and staffed federal regulators had scarcely kept pace with technological developments in the industry. It is doubtful that BP's federally approved oil spill response plan was even read by those who approved it: "BP had named Peter Lutz as a wildlife expert on whom it would rely; he had died several years before BP submitted its plan. BP listed seals and walruses as two species of concern in case of an oil spill in the Gulf; these species never see Gulf waters. And a link in the plan that purported to go to the Marine Spill Response Corporation website actually led to a Japanese entertainment site."[48] UC Berkeley's Robert Bea attributed the 2010 Gulf of Mexico oil spill to a "cascade of poor decisions involving poor tradeoffs made by all of the organizations with primary responsibilities for the Macondo well-project."[49] Although the Deepwater Horizon disaster involved flawed decision-making by specific people and organizations, the implications for effective risk management are widely applicable. Responses to recent catastrophes such as the 2010 earthquake in Haiti, the 2011 Fukushima tsunami and nuclear disaster, and Hurricane Sandy in 2013 reveal unanticipated consequences and weak management arising from complex interactions within and between different kinds of systems.[50]

In his analysis of the potential fragility of all complex engineered systems, Venkat Venkatasubramanian writes: "Modern technological advances are creating a rapidly increasing number of complex engineered systems, processes, and products, which pose considerable challenges in ensuring their proper design, analysis, control, safety, and management for successful operation over their life cycles. It is their scale, nonlinearities, interconnectedness, and interactions with humans and the environment that can make these systems-of-systems fragile, when the cumulative effects of multiple abnormalities can propagate in numerous ways to cause systemic failures."[51] Venkatasubramanian is describing a *systemic action problem* arising from the embedding of increasingly complex, potent, and potentially destructive engineered systems in vulnerable human and environmental systems. A lesson to be mindful of is that as human beings make these systems more complex and potent they are amplifying the challenges of management dramatized in the language of *wickedness*. What had seemed as straightforward as going to the moon turns out to be as systemically daunting as eliminating poverty.

The Future of Fossil Fuels
The 2010 Gulf of Mexico oil spill delivered an unwelcome lesson in the hazards of underestimating the risks associated with drilling in extreme environments. Yet five years later the number of deep-water rigs in the Gulf of Mexico has risen from thirty-five to forty-eight, the oil and gas industry is pushing to drill to even greater depths, and the Obama administration has granted Royal Dutch Shell permission to develop oil fields in the Arctic's Chukchi Sea, one of the world's most hazardous places to drill.[52] Science has established that anthropogenic global warming is real, mostly caused by human activities, and already having impacts. The most recent report of the Intergovernmental Panel on Climate Change warns that "continued emission of greenhouse gases will cause further warming and long-lasting changes in all components of the climate system, increasing the likelihood of severe, pervasive and irreversible impacts for people and ecosystems."[53] Atmospheric carbon dioxide levels meanwhile have continued to reach new historic highs with each passing year.[54] A recent report from the International Energy Agency predicts that fossil fuels will nevertheless continue to dominate world energy until 2040.[55]

These disturbing juxtapositions bring us back to our preface to this case study and why societies court disaster despite their knowledge of the risks. Considered from this wider perspective, the Deepwater Horizon and Texas City cases are not just lessons in organizational epistemic failure and the hazards of complexity, but lessons about the quality of work and prospects for sustainability. Workers who report an exceptional degree of fear of dying in a lethal accident at their place of employment but continue to work there evidently perceive themselves as lacking better opportunities. Those who worked on a "nightmare well which has everyone all over the place" and went home to families and communities that depended on income from both the oil industry and fishing or tourism may have experienced a similar lack of opportunity, even when faced with the painful recognition of the spill's impact on those around them. This is neither a happy state of affairs nor one favorable to sustainability. People who need jobs in fossil fuel industries so much that they would endure exceptional fear of dying on the job could scarcely be faulted for joining their employers in opposing stronger safety regulations and pricing of carbon emissions. Despite what they know firsthand, they might take it for granted that the only bad governance is in the public sphere.

As we have noted before, the terms of an effective global climate accord would need to be implemented through domestic enabling legislation in each signatory state, so perceptions of the fairness of those terms will matter not only globally but also domestically. We suggested a guiding conception of such fairness in the previous two chapters, calling for a reconceptualization of equal opportunity, meaningful education for those who do not go to college, and a distribution of the governance of workplaces that gives workers a stronger voice in the conditions of their employment. In the final analysis, having only one opportunity to live a good life is *not* an opportunity to live well if one's work destroys the opportunities of one's neighbors or one's grandchildren. We cataloged many obstacles to sustainability in chapter 2, from cultural inertia and excessive faith in technology and markets to counterproductive government subsidies for the fossil fuel industry and short corporate and political time horizons for weighing the consequences of decisions. The closing lesson of this case study is that a central focus of advancing sustainability must be to provide everyone with work that is both good for them and does not destroy the opportunities of others to live well—work that satisfies our eudaimonic and preservation of opportunity principles.

Water Governance in the Murray-Darling Basin of Australia

If there is magic on this planet, it is contained in water.
—Loren Eisley

It would be hard to overstate the importance of water. It is an irreplaceable necessity, essential to human well-being in a variety of ways, and essential to the functioning of ecosystems. Water sustains life and livelihoods, is essential for health, and plays vital roles in food and energy systems. We noted in chapter 1 that water scarcity is already evident in "a number of the world's major river basins and it is estimated that by 2030, nearly 50 percent of the global population will be living in areas of high water stress."[56] The quality of the world's freshwater resources is also rapidly declining in many places, and climate change is making matters worse by accelerating the global water cycle, changing the timing, intensity, and patterns of precipitation.[57] These combined threats of excessive water withdrawals and climate change are already seriously compromising the ecological health of several of the world's major river systems, including the Colorado, Yellow, Murray, and Orange Rivers, located in the Western United States, China, Australia, and South Africa, respectively.[58]

During the past century, local and regional water management were the norm, and they relied on engineering-focused approaches in which dams, reservoirs, and other structures were constructed to control the flow and storage of water. Management of water systems was rarely part of an integrated approach that considered the systemic relationships between water, food, and energy. In contrast, there is now growing recognition that the scale and complexity of water-related challenges call for planning and cooperation beyond local and regional levels and that technology and engineering alone cannot solve persistent and growing problems of water availability and quality.[59] Emerging approaches to water management are embracing a more comprehensive and collaborative process based on principles of environmental sustainability and analysis of trade-offs between competing social, economic, and environmental demands for freshwater resources.[60] These approaches demonstrate progress in understanding and managing systemic action problems.

We will focus on Australia's Murray-Darling Basin, where water reform has been a decades-long undertaking for one of the world's largest rivers and its driest continent. By the 1990s, it was apparent that existing approaches to dividing limited water between irrigation and other uses were leading to overallocation and significant damage to the Murray-Darling's ecological and hydrological systems. The record-breaking Millennium Drought of 1997–2009 exacerbated existing environmental problems and prompted a series of ambitious water policy reforms, which, though not without problems, have established Australia as a world leader in adaptive water management.[61]

Historical Overview

As residents of the world's driest inhabited landmass, the people of Australia have a long and complicated relationship with water. Because Australia not only is dry but possesses one of the world's most variable rainfall patterns, attempts to cope with extremes in water availability are woven throughout Australia's history and are part of its national identity.[62] Australia's water problems are in part a reflection of its geography. It is located in latitudes characterized by low precipitation, and El Niño, La Niña, and other weather patterns in the South Pacific Ocean contribute to boom or bust rainfall. Australia also experiences high variation in rainfall from region to region, with 25 percent of its landmass receiving 75 percent of its precipitation.[63]

From the time of the First Fleet's arrival in Sydney Harbor in 1788, European settlers devoted themselves to "fixing" Australia's water problems. By 1945, the country had built most of its more than five hundred dams, and from then until the mid-1960s it tripled the extent of its irrigated land. By the 1980s, Australia's waterways were among the most regulated in the world, and some questioned whether the Murray River, Australia's largest, even qualified as a river anymore.[64] According to R. Quentin Grafton and his coauthors, the most rapid expansion of irrigation took place during a relatively wet period lasting from the 1950s to the 1980s and did not take into account climate variability and the need to plan for the next drought. Although further increases in water use for irrigation were capped in 1995, it took the Millennium Drought, which produced the lowest average net inflows into the Murray River ever recorded during a five-year period, to force a reduction in the amount of water diverted for agriculture.[65]

One of the largest river systems in the world, the Murray-Darling River Basin covers more than one million square kilometers of inland southeast Australia and passes through five states and territories.[66] The basin is notable as one of the most variable river systems on Earth with respect to rainfall and stream flows.[67] More than two million people, including about thirty Aboriginal nations, rely on the basin, which contains some of the nation's most valuable environmental and agricultural land, known as Australia's "food bowl." A number of challenges in the transboundary Murray-Darling Basin arise from the fact that water management historically has been primarily a state function in Australia. Responsibility for water resource management historically has been divided among a wide range of state, local, and federal institutions, and adoption of widely different policies regarding water entitlement and allocations by the different states led in

time to overallocation of limited water resources and significant environmental deterioration.[68]

Attempts to improve the management of the Murray-Darling Basin
began in 1863, when the first planning conference was held. Many conferences and conventions followed, and in 1915, prompted by the severe Fed-
eration Drought of 1895–1901, the governments of New South Wales,
Victoria, South Australia, and the Commonwealth (Australia's federal government) signed the River Murray Waters Agreement (RMWA). This was
followed in 1917 by establishment of the River Murray Convention to coordinate more efficient sharing of the river's water. The following decades
were marked by construction of pipelines, tunnels, and dams to support
irrigated agriculture and generate hydroelectricity, and the 1917 agreement
was amended several times in response to changing community priorities
and economic conditions. To promote irrigated agriculture, water entitlements generally were free during this period.

By the late 1960s, poor irrigation practices led to water quality problems—notably, rising salinity levels—and in 1982 and 1984 the RMWA was
amended to address environmental degradation. By the early 1990s, it was
apparent that these basin-wide problems could not be adequately addressed
by individual states with discrete water-management strategies, and in
1992 the Murray-Darling Basin Agreement was signed by New South Wales,
Victoria, South Australia, and the Commonwealth. In spite of this new
agreement, conditions in the basin continued to worsen.[69] From 1991 to
1992, the world's largest recorded blue-green algal bloom spread for more
than 1,000 km along the Barwon-Darling River. A state of emergency was
declared, drinking water supplies had to be brought in, and some livestock
died from drinking contaminated water.[70] The inadequacies of existing
approaches to water management had become clear.

In 1994, the focus of water policy shifted to market-based reforms, which
included trading in water allocations or entitlements.[71] This process served
to separate the right to use water from land ownership through issuing
water licenses or *entitlements* that could be bought and sold.[72] In 1995, a cap
was set on surface water diversions, and issuance of new water rights in the
basin was suspended. In spite of these and subsequent reforms, the Millennium Drought heightened concerns that existing policies were failing to
preserve the environmental health of the Murray-Darling Basin. A 2004
National Water Initiative advanced a new policy agenda building on earlier
reforms, and the 2007 Water Act shifted management of the basin from the
state governments that had controlled it since 1901 to the Australian government.[73] The newly created Murray-Darling Basin Authority (MDBA) was

charged with developing a plan that would protect the environment by setting a *sustainable diversion limit* designed to control the amount of water that could be extracted from rivers and groundwater systems across the basin while optimizing social and economic outcomes.[74]

In 2010, MDBA released its "Guide to the Proposed Murray-Darling Basin Plan," which called for three thousand to four thousand gigaliters of water (around 30 percent of total water use in the basin) to be returned to the environment. This plan was vehemently protested by farmers, and intense media coverage of contentious exchanges about the proposed plan ensued.[75] After changes in leadership, revisions of the plan in response to commentary, and intense political debate, Basin Plan 2012 was signed into law in November 2012.

Water markets, which ended perceptions of water as an unlimited and cost-free resource, are a critical part of Basin Plan 2012 and previous water reform efforts in Australia. Other strategies in Australia's water reform tool kit include modernization of irrigation infrastructure to improve water delivery and creation of entitlements for the environment (i.e., allocations of water to be returned to or remain in the environment) in order to protect ecosystem health and capacity. The power of water markets is illustrated by statistical data compiled during the Millennium Drought, which show that despite a decline in irrigated water use by about 70 percent, the nominal gross value of agriculture fell by less than 1 percent, due in part to shifts to higher-value crops and horticulture facilitated by water trading.[76] In cities, strategies were employed to increase supply and reduce the use of water. Expensive water recycling and desalination plants were built, and water-use restrictions ranged from prohibition of car washing and daytime watering to rules promoting more efficient use of water. Per capita water use in urban areas decreased by 50 percent from 2002 to 2009, a trend that continued beyond the drought.[77]

Adrian Piani has identified several key ingredients for success in the Murray-Darling Basin Plan: (1) the guidance of strong leaders who are able to communicate clearly, foster a strong spirit of collaboration and ownership, and understand the value of talking directly with stakeholders; (2) water trade and the entitlement process; (3) funding (the Australian Government has invested billions of dollars in the reform efforts); and (4) scientific input and state-of-the art modeling to assess current and projected environmental water needs in a large and complex river system.[78]

Australia's water reforms are also notable for their built-in flexibility. Although the sustainable diversion limit is predetermined, the Basin Plan 2012 plan provides flexibility in how goals can be met, and governments can propose projects that deliver equivalent environmental protections with less water as long as other outcomes are not compromised in the process. This built-in flexibility is designed to minimize risks associated with future drought by enhancing resilience within the basin's environmental systems.[79]

Progress and Pitfalls

Implementation of the Murray-Darling Basin Plan is a work in progress and there have been significant hurdles to the implementation of water markets in Australia, including the costs of acquiring the requisite information for establishing effective trading rules, state self-interest, concern about the loss of water in some communities, fear that "water barons" might emerge, and water-management personnel with backgrounds in science and engineering but no experience in market-based mechanisms.[80] The National Water Commission Act of 2004 and its amendments require periodic independent and evidence-based assessments of progress toward National Water Initiative (NWI) objectives. The fourth (and final) assessment issued by the National Water Commission in 2014 warns that "there is a real risk of gradual backsliding on current progress, and a retreat from public accountability."[81] The Wentworth Group of Concerned Scientists has echoed these concerns and called for renewed and expanded commitment to reform, citing the 2014 abolition of the National Water Commission and the Council of Australian Government's Standing Council on Water and Environment and a weakening of programs designed to return water to the environment.[82] Australia's systemic water reforms were enacted in response to an unprecedented drought that magnified longstanding problems related to water scarcity and environmental degradation. Yet, the sense of urgency inspired by the crisis was undermined in 2011 by heavy rains and widespread flooding within the Murray-Darling Basin.[83] These conditions have not reduced the long-term risk for which Australia must prepare, but it is harder to enlist cooperation in preparing for future risk in the absence of an immediate crisis.[84]

Water markets are the centerpiece of Australia's water reform, and they have been employed in other countries as well, including China, Chile, South Africa, and the Western United States. Although water trading has proven beneficial in some contexts by facilitating more efficient and flexible allocation of oversubscribed water resources, water commercialization or the *ownership* of water by for-profit corporations has given rise to serious problems.[85] Water trading benefits those engaged in the transaction but may have adverse impacts on others, including indigenous groups and those whose livelihoods depend on the noncommercial benefits of ecosystems.[86] Maude Barlow, who was instrumental in the United Nations' recognition of water for basic personal necessities as a universal human right, refers to the commercialization of water as a "dangerous new trend in the commodification of water," which takes control of Australia's scare water supplies away from its citizens. She argues that Australia's water laws permitting anyone, not just smaller agricultural users, to buy water entitlements have led to skyrocketing costs, making it difficult for the government to buy back water for the environment and creating a lucrative market that favors big business and other private interests.[87] When water that has been held as a public trust is bought by multinational corporations, the bargaining position of states and municipalities may change dramatically for the worse. Legitimate concerns about the drawbacks of commercializing water do not preclude water trading, but they suggest that private water ownership and markets should be limited and structured to protect natural water systems as public trusts and honor the universal right of affordable access to clean water for drinking, cooking, and personal hygiene that was established as a provision of international law by the United Nations General Assembly and the UN Human Rights Council.[88]

Emerging Strategies

The lessons of the Australian experience should prove helpful in managing other transboundary river basins, of which there are many.[89] In total, 276 river basins are shared by more than one country, and together these cover nearly half of the planet's land surface. Thousands of other rivers flow through two or more jurisdictions within a single nation.[90] There can be no one blueprint for water management applicable everywhere and in all conditions, but Australia's water reform efforts provide an instructive example of innovative water management designed to address the complexities of water allocation in complexly interrelated human and hydrological systems. Across the world, new paradigms, tools, and infrastructures already are being invented to foster a more integrative, collaborative, and adaptive approach to water management. Initiatives and experiments in fostering and researching collective management of watersheds are not only facilitating cooperative management of common water resources but also advancing knowledge of the strengths and limitations of such management.[91]

Nathan Engle and his coauthors describe two of the most influential recent trends in water management: integrated water resources management (IWRM) and adaptive management (AM). IWRM focuses on decentralization of policy and integration of previously isolated aspects of water resource management, including surface and groundwater, water supply and quality, and the interconnected biophysical and socioeconomic dimensions of water management. IWRM aims to integrate strategies and disparate stakeholder interests across multiple scales to inform inclusive decision-making. AM is chiefly concerned with managing uncertainty through experimentation and monitoring feedback from the affected socio-hydrologic systems to guide next steps.[92] There has been a recent trend to merge these two approaches to better address the growing complexity of water management, but little empirical evidence exists for assessing the potential effectiveness of these strategies in practice.[93]

The literature is replete with references to the failure of traditional command-and-control approaches to adequately address complex socioecological interactions in transboundary basins. Yet many researchers caution that IWRM, AM, and related approaches are still difficult to apply and that there is much to learn about collaborative management of resource systems in different socioecological contexts. As our previous case study suggests, the involvement of many stakeholders in the governance process also adds institutional complexity that may hinder the effective integration of diverse perspectives.[94]

The importance of science for water management has long been recognized, and many river basin plans (including Australia's Murray-Darling Basin Plan) advocate the use of the best-available science—yet perspectives are changing.[95] As we saw in chapter 4, there is wide and growing recognition that scientific knowledge and methods are essential but not sufficient and that it is important to overcome the divide between scientists and representatives of the public interest. While acknowledging the difficulty of combining science with other perspectives important to managing transboundary resource system problems, Derek Armitage and his colleagues emphasize the potential benefits: "The increased participation by a greater array of non-government actors in transboundary settings can lead to greater legitimacy, more effective and equitable allocation of resources, a better ratio of costs to benefits, and an improved access to a diversity of knowledge and expertise ... as well as broader acceptance and implementation success."[96] The transboundary resource system problems under discussion qualify as systemic action problems, as does our final case: the food and water systems of the Mekong River, which flows through six nations.

Food and Farming in the Mekong Region of Southeast Asia

Tell me what you eat and I'll tell you who you are.
—Jean Anthelme Brillat-Savarin[97]

Food has long been as much an expression of identity, culture, and status as a staple of life, and dietary preferences play a significant role in the carbon emissions that are largely responsible for climate destabilization. The Greenland Norse's disdain for the fish they easily could have survived on was an aspect of their European identity and favorable self-regard, and exclusive dining remains a coveted status symbol in many circles.[98] Greater body mass was such a conspicuous marker of class privilege and obstacle to social unity and military efficiency in ancient Sparta that it was legislated out of existence; a system of common meals for all Spartan men ensured they would all live and fight together on the same diet.[99] Such measures are all but unthinkable in the contemporary world, but starvation is as much a threat today as it was for the poor of ancient Sparta. Food of sufficient quantity and quality is a fundamental prerequisite for living well—and one that societies must be able to provide their members if they are to safeguard future opportunity. Food security remains an elusive goal, however, and the dietary fate of many countries lies substantially beyond their control. It is here and in water security above all that the goals of development and sustainability must be carefully aligned.

Today's food systems are complex, ranging in scale from the local to the global and encompassing both traditional farming and industrialized agriculture.[100] Despite decades of significant gains in agricultural productivity through the use of high-yield seeds, fossil fuel–based fertilizers and pesticides, and extensive irrigation and mechanization—all aspects of the Green Revolution—many consider the contemporary global food system to be at a turning point. The evidence that it is faltering is substantial and particularly worrisome in light of the ongoing desertification, declining fisheries, deforestation, and climate change reviewed in chapter 1. Nearly a billion people in developing countries are undernourished, while another billion people are overweight, and nearly half a billion suffer from obesity, creating a double burden of obesity and hunger in affluent and poor countries alike.[101] Compounding these concerns, modern agriculture, with all its attendant greenhouse gas emissions, land-use changes, biodiversity loss, disruption of the nitrogen cycle, demands on fresh water supplies, and polluting pesticides and fertilizers, constitutes what Jeffrey Sachs has called the "single largest source of human-induced environmental change."[102]

Behind these aggregated statistics, there are of course local and regional contexts and conditions important to managing and mitigating the problems.[103] The region we will consider briefly is Southeast Asia, where 232 million people are undernourished and policy makers struggle with numerous and varied threats to future food supplies.[104] The Mekong Basin is an extensive and populous area encompassing the territories of six very different countries connected by the world's ninth largest river. One of its notable features is that it is a region of food production that is undergoing rapid economic growth—growth that will inevitably involve trade-offs and conflicts in allocating interdependent water, food, and energy resources. An important focus of growth, trade-offs, and conflicts is hydropower development to meet growing demands for energy and the threat it poses to essential ecosystem services that provide food security and livelihoods for millions of people.[105] The same dams that are expected to support improved agricultural productivity though enhanced irrigation will also adversely impact food production in downstream countries by altering natural stream flows, sediment transport, nutrient cycling and fish migration.[106]

More than seventy-two million people live in the eight hundred thousand square kilometer basin of the Mekong River, which traverses six countries along its approximately five-thousand-kilometer path from Tibet to the South China Sea. As a transnational river, the Mekong connects many and diverse stakeholders with different perspectives, values, goals, and governments. People have relied on the rich and varied resources of this region for thousands of years, and food in the wider Mekong area reflects the ecological diversity of the regions traversed by the river, from the farmed and forested uplands to lowland plains and plateaus to Cambodia's Tonlé Sap to coastal areas and deltas. These varied terrains yield a wide array of products, including rice, fruit, legumes, and livestock.[107] Today, the region is dominated by the large economies of Thailand, Vietnam, and China, in addition to the three lesser-developed nations of Cambodia, Laos, and Myanmar.[108]

The wider Mekong Region is experiencing rapid economic, social, and environmental changes that have implications not only for the future of agriculture and food systems within the region itself, but for the world. The region serves as the world's "rice bowl." It is home to the world's largest inland fishery and is a biodiversity hotspot second only to the Amazon region. The changes in the region are evident in its growing population and demand for food; a shift from traditional subsistence farming to commercial agriculture; urbanization; dietary changes favoring meat and other resource-intensive foods; increasing demands for freshwater needed for agriculture, cities, and industry; and construction of dams for expanded irrigation and hydropower. The uncertain and unevenly distributed impact of climate change, including warming, changing rainfall patterns, and rising sea levels, will further affect food production in the region.[109]

Green Revolution technologies have increased agricultural production in the Mekong Region through the past two decades, but food security remains weak in rural areas.[110] The gains in production have come at a steep environmental cost in the form of water pollution, land degradation, and loss of forests and biodiversity. Dams and other large-scale infrastructure projects will diminish natural resources vital to the livelihoods of the rural poor, though aggregate regional increases in economic wealth obscure the attendant loss of opportunity for millions of individuals.[111]

As with other transnational river basins, the relationship between upstream and downstream stakeholders is marked by competing priorities. For example, Laos wants to develop its hydropower potential; Thailand seeks cheap hydropower and more water for its industrialized agricultural system; Cambodia prefers to preserve the current hydrological regime, including the seasonal flooding that supports its fisheries; and Vietnam wishes to protect its agriculture and aquaculture from the effects of rising sea levels. These competing national needs and aspirations contribute to the complexity of managing the cross-boundary waters of the Mekong River. Political leaders in the region understand that their countries' futures are intertwined and will depend substantially on well-coordinated water management.[112] However, this understanding is not enough to ensure equitable cooperation. The various leaders are not likely to have comparable influence over the outcome of negotiations, because the bargaining position of countries is often influenced by differences in development and wealth, political influence, and geography.[113] Because rivers always flow from higher to lower elevations, there are inevitability upstream and downstream players, with the upstream players usually getting their way.[114] Competing national interests and complex histories of power relations among neighboring Mekong nations still influence their interactions and pose significant obstacles to coordinated river management.[115]

Recognition of the need for cooperation across the Mekong Basin began in the late 1950s, but the Cold War and regional conflicts ruined any prospects for cooperation until Cambodia's civil war ended in the early 1990s. In 1995, the four countries of the Lower Mekong Basin (Cambodia, Laos, Thailand, and Vietnam) committed to the 1995 Mekong Agreement through the creation of the Mekong River Commission (MRC).[116] Despite this, there has been little progress in overcoming a

fragmentation of governance attributable to conflicting national interests.[117] This fragmentation is most apparent in the conflict surrounding
construction of dams for hydropower, an activity that disproportionately
affects the estimated 60 million people who rely on wild-caught fish from
the Mekong River as their major source of protein and nutrients. In spite of
potentially devastating losses of food and biodiversity, plans for dam construction proceed.[118]

The Mekong Agreement requires that any Lower Mekong country proposing a dam project to further its own interests must first consult with the
other MRC states.[119] The first major test of the Mekong Agreement came
with construction of the Xayaburi dam, one of several dams that Laos plans
to build on the mainstream Mekong. Instead of fully cooperating with the
governments of the other Lower Mekong countries, Laos proceeded with
the project in spite of the significant concerns expressed by Cambodia and
Vietnam. Meanwhile, Thailand quietly financed the project and agreed to
purchase electricity generated by the dam. According to Kirk Herbertson,
Laos's failure to fully cooperate with other countries around the Xayaburi
dam project set a precedent that threatens to undermine future attempts at
cooperation under the auspices of the Mekong Agreement.[120]

In addition to dam-related threats to food security, a drought and
increased saltwater intrusion are threatening agriculture in coastal areas.[121]
In spring 2016, the worst drought in Vietnam's recorded history jeopar-
dized key crops such as rice, cassava, and maize. Making matters worse,
nearly a million people in Vietnam had no access to fresh drinking water.
Low water levels typical of a normal dry season cause saltwater from the
South China Sea to flow into the Mekong, but the drought caused such
saltwater intrusion to begin earlier, contaminating rice paddies and groundwater supplies as far as ninety kilometers inland. With warmer temperatures and rising sea levels due to climate change, there is concern that
coastal areas of the Mekong region will become unsuitable for cultivating
some traditional subsistence crops.[122]

Despite considerable opposition, development in the Greater Mekong
continues to be dominated by construction of dams for hydropower at the
expense of food security. The benefits and costs of energy production are
unevenly distributed between countries and within countries, and the
patterns of distribution generally favor those with greater influence and
bargaining position: "The Mekong region and its 'exceptional untapped
potential' is seen as 'ripe' for massive investments in hydropower, flood
control and irrigation infrastructures. On the other hand, civil society
groups operate with a more critical worldview, which emphasizes the social

and environmental costs of transformations, and how they overwhelmingly benefit political or economic elites. Current project planning and implementation in the region tend to confirm that decision-making processes are often opaque and offer limited support to the claim that 'we have learned from past mistakes.'"[123] Here as elsewhere, there may nevertheless be hope that the costs of failing to cooperate will sooner or later fall on everyone and that "a culture of negotiation and social learning" might be cultivated in the context of governance structures that are more effectively and prudently collaborative, comprehensive, and adaptive.[124]

Conclusion

Consideration of the structure of sustainability problems led us to identify them as *systemic action problems* in the opening section of this chapter, and we went on to consider both in general terms and through case studies how such problems can be managed in the face of challenges to societies' collective understanding, capabilities, and virtues of cooperation. We approached this analysis, as much of the literature of sustainability does, from a management perspective while suggesting that the strengths of collective action must be nurtured and orchestrated by a political art and system of just institutions that rely on distributions of governance sensitive to both the requirements of regional and global coordination and the limitations of centralized and top-down management. The movement toward more collaborative, comprehensive, and adaptive management noted in this chapter is a promising start toward creating such institutions, and it is through ongoing participation in the collaborative governance mediated by such institutions that we would expect much of the needed learning and capacity building to occur.

In the final chapter of this book, we will take a longer view of such learning and capacity building. We address the educational means by which societies can lay a broad foundation of understanding, capabilities, and virtues conducive to sustainability. In doing so, we will pick up the threads of our discussion of just educational institutions and opportunity begun in chapters 3 and 4.

6 An Education in Sustainability

Plato arrives at his proposed curriculum for a system of public education in the seventh book of his *Laws*, having elaborated and defended his vision of a constitutional rule of law imposed not by force but through the rational consent of citizens who share in governing—a system designed to enable all its citizens to live well now and in the future. In the words of the dialogue's Athenian Stranger, Plato writes: "When I look back over this discussion of ours ... it's *these* [collected chapters] that have impressed me as being the most eminently acceptable and most entirely appropriate for the ears of the younger generation."[1] Without feigning the immense satisfaction Plato's Athenian Stranger expresses, we will begin this concluding chapter by acknowledging that the foregoing chapters on the nature, ethics, and pursuit of sustainability are both a curriculum of sustainability and a basis for concluding that an education in sustainability is a requirement of justice. Just educational institutions should provide children with an education in sustainability, and the principal task of this chapter is to explain why and to outline and defend the education in sustainability we propose.

In chapter 2, we identified deficiencies of education as an obstacle to sustainability, but we also identified ways in which suitable education might be helpful in overcoming other obstacles. We noted the potential value of education in facilitating the understanding, communication, shared norms, trust, and leadership helpful to local and regional self-organization to protect environmental commons. We also noted the relevance of critical thinking coupled with understanding of environmental and social systems and the importance of universal elementary education for reproductive freedom and population stability. Deficiencies in systems of public knowledge call for education that develops stronger attunement to scientific evidence and explanations and a critical understanding of advertising and related media—an understanding often referred to as *media literacy*. Our discussion of the psychosocial dynamics of denial also made it clear that schools must be communities of thought and practice that offer students constructive outlets for the troubling emotions associated with learning about climate change and other problems of sustainability—opportunities to engage in actions consistent with a sustainable future. Chapter 3 followed with an ethical framework and theory of justice that should be part of the content of an education in sustainability, presented not as doctrine but for rational examination. Examining what it is to live well and what can be learned from related research on motivation, psychological needs, and well-being is a natural part of an education in sustainability. Chapter 4 then outlined a philosophy of education that can ground education in sustainability and diagnosed some limitations of educational systems that might be overcome through reforms favorable to sustainability. Focusing on dynamics of complexity and sociopolitical collapse, chapter 4 paved the way for addressing the management of systemic action problems and the water-food-energy nexus in chapter 5.

A curriculum based on these and related starting points in the preceding chapters would include the following:

• Initiation into scientific thinking about complex systems, with explicit attention to the nature of scientific inquiry, evidence, and explanation
• Geology, ocean and climate science, ecology, and the history of life
• Geography, socioecological systems, and patterns of societal survival and collapse
• Economic and political world history, with attention to resources and production, energy transitions, environmental impact and governance, development policy, debt, and institutional platforms for international cooperation
• Local and global citizenship and cooperation
• Technology and creative design
• Lessons and practice in critical and creative thinking, focused on the diagnosis and remediation of failures of rationality, imagination, and foresight in responding to evidence and uncertainty, anticipating consequences, and contemplating alternatives
• Media literacy and practice in thinking critically about advertising, culture, and life choices
• Psychology and health, with an emphasis on well-being, emotion, and motivation
• Sustainability ethics

General education along these lines would have a greater impact if it were integrated with occupational and professional training through the duration of such training. For many students, such training could begin in school rather than college, as we suggested in chapter 4. The impact of an education in sustainability could be much more immediate and widespread in a system that allowed more students to put their sustainability-related learning into practice at work and in lives that are not burdened with the insecurities of long and uncertain transitions to adulthood.

How close do schools come to providing an education in sustainability (EiS, ESD, or EFS) that combines these elements?[2] To what extent are the starting points or foundations for such an education in place? What is the policy context for advancing an education in sustainability in schools in North America and in the United States in particular? We will begin by answering these questions, considering UNESCO's Decade of Education for Sustainable Development (DESD, 2005–2014), the progress toward implementing its goals in some countries, and the notable barriers to education in sustainability in the United States. Education for sustainable development (ESD) is a vision of how to address problems of sustainability and poverty, predicated on the assumptions about their relationships to one another noted in chapter 1. Various criticisms of ESD will be noted: Is it a well-defined, coherent package? Is it too prescriptive? Does it incorporate an appropriate form of environmental education? Is it based in a sound conception of education? We will answer these questions as a prelude to our own conception of education in sustainability (EiS).

Why is EiS warranted? The simplest and most compelling reason for providing EiS is that everyone is entitled to it. Young people are entitled to education that offers them substantial opportunities to live well in the world in which they will live, which for the foreseeable future will be a world of increasing ecological and societal risk. A second reason for providing such education is that there are reasons of both morality and prudence for cooperating to solve the problems of sustainability, and these reasons for cooperating make the form of EiS we advocate desirable. It follows that EiS is not just educationally legitimate but essential to an adequate education. We will expand upon these arguments and offer a vision of EiS grounded in our conception of just educational institutions.

UNESCO's Decade of Education for Sustainable Development

The term *sustainable development* is usually defined as it was in 1987 by the World Commission on Environment and Development (WCED), also known as the Brundlandt Commission: "development that meets the needs of the present without compromising the ability of future generations to meet their own needs" or, somewhat more expansively, "development that meets the needs *and aspirations* of the present without compromising the ability to meet those in the future."[3] Thus defined, the primary focus of sustainable development is on development "to ensure steady improvement in the quality of life," with the understanding that such improvement must continue for "future generations" and "in a way that respects our common heritage—the planet we live on."[4] The United Nations Educational, Scientific and Cultural Organization (UNESCO) later endorsed a definition of sustainable development put forward in a jointly sponsored 1991 document entitled *Caring for the Earth: A Strategy for Sustainable Living.* There, the term is defined as "improving the quality of human life while living within the carrying capacity of supporting ecosystems."[5] Education for sustainable development has related roots in development education and is not simply an extension of environmental education.[6]

In December 2002, the United Nations General Assembly resolved to declare and implement a United Nations Decade of Education for Sustainable Development from 2005 to 2014.[7] UNESCO took the lead in this effort, enlisted an array of institutional and research partners, and developed a web-based multimedia teacher education program (Teaching and Learning for a Sustainable Future) and an International Implementation Scheme (IIS). On the basis of extensive consultations, the scheme was developed and presented to the UN General Assembly in October 2004 and to UNESCO's executive board in April and September 2005. The IIS set out a framework for implementation, premised on global collective ownership of the DESD. To this end, it provided a rationale and identified strategies and milestones while also relying on partnerships with cooperating nations and links to "other key education movements."[8] The primary goal of the DESD was to encourage governments to "consider the inclusion ... of measures to implement the Decade in their respective education systems ... and to promote public awareness of and wider participation in the Decade."[9]

The *educational vision* promoted by the DESD has five components:

1.   Universal access to adequate basic education
2.   Reorientation of *all* education programs to address sustainable development
3.   Promoting wide public awareness and understanding of sustainability, through not only schools but also wider public education campaigns
4.   Training in the skills—technical, analytical, and social—required for "social, economic, and environmental sustainability"

5. Involvement of higher education in sustainability research, learning, and DESD implementation[10]

Substantively, the education to be promoted is described as resting on four *pillars*: "recognition of the challenge," "collective responsibility and constructive partnership," "acting with determination," and "the indivisibility of human dignity." These pillars together imply

• scientific understanding of environmental, resource, and climate problems;
• the scientific, technical, and critical thinking "tools" needed to "transform current societies to more sustainable societies"; and
• "the values, behavior and lifestyles required for a sustainable future."[11]

The *environmental education* component is focused on the natural resources and ecosystem goods and services "essential for human development and indeed survival," and *environmental literacy* is understood broadly to include "the capacity to identify root causes of threats to sustainable development and the values, motivations and skills to address them."[12]

The *values* to be taught are regarded as uncontroversial:

• Respect for diverse peoples and cultures
• Nonaggression
• Commitment to global partnership in living sustainably and raising living standards for the worst off

The DESD's educational vision thus incorporates a form of values education that is grounded in an objective understanding of resource depletion and ecological risk but it is also built on assumptions about the desirability or moral necessity of responsible global citizenship, a peaceable and just approach to achieving a collectively sustainable ecological footprint, and intergenerational equity.

ESD is to be interdisciplinary, values-driven, grounded in integrative environmental science, and focused on developing the critical thinking and problem-solving skills learners will need to confidently address "the dilemmas and challenges of sustainable development."[13] The UNDESD does not provide a detailed model of ESD, but rather a rationale and broad conception of its essential features, together with a web-based environmental studies (environmental literacy) resource for teachers.[14] Local initiative and choice in working out the details of ESD are accepted as critical to success. This is to be expected, given the importance of adapting education to local circumstances and the limitations of UNESCO authority and resources.

ESD Implementation

The DESD was announced with a sense of urgency evident in its references to *human survival*, and it was taken seriously in the United Kingdom, Germany and elsewhere in Europe, Australia, and Japan, with national governments playing major roles in implementation through curriculum standards, allocation of resources, sponsorship of research on best practices, and assessment schemes.[15] The pace and extent of implementation was and remains problematic, however.[16] In North America, it has been even slower and more fragmentary. This is explained to some extent by the fact that Canada and the United States both lack federal systems of education and national curricula. In Canada, authority over learning standards rests in ten provinces and three territories, and in the United States authority rests in the fifty states. Both countries have also experienced a narrowing of focus on reading and mathematics and the influence of related testing that discourages expenditure of classroom time on critical thinking and interdisciplinary instruction essential to education in sustainability. As a result, progress toward implementing any form of sustainability education has been slow in both countries and is typically led by individual provinces or states or by highly committed teachers and administrators.[17] In the United States, the DESD is barely discussed, and educational developments related to sustainability are mostly confined to higher education and focused on sustainable campuses.[18]

Shaping the Future We Want, the final report of the DESD, concludes that at the end of ten years "a solid foundation has been laid for Education for Sustainable Development achieved by raising awareness, influencing policies and generating significant numbers of good practice projects in all areas and levels of education."[19] The report also acknowledges that in spite of progress, the decade did not achieve a complete transformation of education systems to fully embrace ESD. In December 2014 UNESCO issued a *Roadmap for Implementing the Global Action Programme on Education for Sustainable Development*, which seeks to maintain the momentum of the DESD by proposing strategies to scale up successful projects, build capacity for more systemic impact, and monitor progress.[20]

The US National Context

In response to UNESCO's roadmap, the US ESD Delegation to the UNESCO World Conference on ESD issued the ESD Roadmap and Implementation Recommendations for the United States of America.[21] Another report, *The*

Status of Education for Sustainable Development (ESD) in the United States, supplements the roadmap with a catalog of existing initiatives at the local, regional, and national levels that support the Global Action Programme on ESD goals.[22] In their retrospective on progress made in the United States during the DESD, Debra Rowe, Susan Gentile, and Lilah Clevey recount a number of success stories spanning higher education, the K–12 sector, and informal education, but they note that ESD has yet to be formalized through inclusion in educational standards, assessments of student performance, and commitment of requisite resources.[23]

Challenges

Fundamental challenges to implementing sustainability education in American schools include the following:[24]

• Low levels of awareness among educational policymakers, school and district administrators, and teachers
• Lack of clarity about the concept of sustainability and contested visions of the purpose of education
• An already crowded, discipline-based curriculum
• Scarcity of teacher education in the concepts and pedagogies of sustainability
• A national focus on high-stakes, fact-oriented testing of disciplinary knowledge

ESD requires problem-focused curricular integration, but curricula in the United States are structured around discrete core subjects, and teachers have little time to coordinate an integrated curriculum with colleagues in other subject areas. "Second-tier" subjects, such as art and health, and "adjectival educations," such as environmental and citizenship education, are given short shrift. Another key factor limiting implementation of sustainability education in the United States is a dearth of appropriate teacher-preparation programs and professional development opportunities for practicing teachers and school administrators. Victor Nolet has described the situation, noting that sustainability education is almost nonexistent in the American teacher-education curriculum.[25] Teacher-preparation programs are already overstuffed with state-mandated requirements and have little incentive to hire ESD-competent teacher educators.

Prospects for the Future

Standards-based education and accountability through high-stakes testing have dominated educational policy and practice in the United States for some years now. The standardized testing requirements of the No Child Left Behind (NCLB) Act of 2001 led to a narrow focus on reading and mathematics literacy in many schools, creating an unfavorable climate for the integrated, transdisciplinary learning essential to meaningful sustainability education.[26] The Every Student Succeeds Act (ESSA) of 2015 relaxed the testing requirements of NCLB, and it authorizes new funding for environmental education, but the impact of standardized tests is not likely to disappear.[27]

The current wave of educational reform in the United States is usually framed in terms of workforce development and the need to bolster economic competitiveness, but it is nevertheless well-aligned in several ways with the approach to education envisioned in the DESD. *Education for Life and Work*, a recent report from the National Research Council (NRC), supplies an overview of current shifts in the educational landscape that have the potential to create more favorable conditions for sustainability education. The report concludes that although "Americans have long recognized that investments in public education contribute to the common good, enhancing national prosperity and supporting stable families, neighborhoods, and communities," formidable economic, environmental, and social challenges render education even more critical today.[28] In recognition that the world has changed drastically since incumbent educational structures and policies were developed, business and political leaders are urging schools to develop essential competencies, often referred to as *twenty-first century skills*, which students need to "reach their full potential in their future adult roles as citizens, employees, managers, [and] parents."[29] These essential competencies include "critical thinking, problem solving, collaboration, effective communication, motivation, persistence, and learning to learn."[30] The report also identifies general aspects of the US education system that may "hinder wider implementation of educational interventions to support the process of deeper learning and the development of 21st century competencies," including lack of appropriate teacher preparation and professional development and assessment that focuses on lower-order thinking and skills.[31] Although criticized by some as no more than the latest fad in education reform, the twenty-first-century skills movement is fostering a national dialog about the need to update education to better align with the complexities of the world students will face.

Two sets of new standards that encompass twenty-first-century competencies, the *Common Core State Standards Initiative* (CCSSI) and the *Next Generation Science Standards* (NGSS), will have a nationwide impact on education for the foreseeable future and may help to create a more inviting context for bringing education in sustainability to K–12 schools. Although led by different groups and developed through different pathways, the Common Core and Next Generation Science Standards both build on research about how students learn, and both focus on better preparing students for college, the twenty-first-century workforce, and making decisions as informed global citizens. The new standards represent a fundamental shift from policies that focus chiefly on recall of facts to a deeper, more integrated learning that develops gradually over time.[32]

The Common Core standards address English language arts (ELA) and mathematics. The ELA standards include guides for reading in the core disciplines of science, history, and social studies, and skills related to the use of media and technology are integrated throughout the standards.

The NGSS were developed through a two-step process led by the NRC, the National Science Teachers Association, the American Association for the Advancement of Science, and Achieve (a bipartisan, nonprofit organization).[33] The NRC began by developing *A Framework for K–12 Science Education: Practices, Crosscutting Concepts, and Core Ideas*, which is anchored in the latest research on science and science learning and identifies "what all students should know in preparation for their individual lives and for their roles as citizens in this technology-rich and scientifically complex world."[34] After a period of public commentary, final development of the standards was led by Achieve and based on the framework.[35] The framework and NGSS are based on three integrated "dimensions" (Disciplinary Core Ideas, Scientific and Engineering Practices, and Cross-Cutting Concepts), which together present a coherent vision of science as not just a body of knowledge that captures our current understanding of the world, but also as a set of practices "used to establish, extend, and refine that knowledge."[36]

The Cross-Cutting Concepts featured in the NGSS (patterns; cause and effect: mechanism and explanation; scale, proportion and quantity; systems and system models; energy and matter; flows, cycles, and conservation; structure and function; and stability and change) are particularly relevant to sustainability education in that they bridge disciplinary boundaries and "help provide students with an organizational framework for connecting knowledge from the various disciplines into a coherent and scientifically based view of the world."[37] As noted by Michael Wysession, the Next Generation Science Standards not only emphasize skills that help students to comprehend, analyze, and apply sustainability principles, they also feature content central to education in sustainability, including a greater emphasis on climate change, an Earth systems science approach, and explicit attention to interactions among human and natural systems. This is reflected in the inclusion of human impacts (middle school) and human sustainability (high school) within the Earth and space science standards.[38] We note, however, the cautions offered by Noah Feinstein and Kathryn Kirchgasler in their analysis of *how* sustainability is incorporated in the NGSS. They write that the NGSS represent sustainability as "a set of global problems affecting all humans equally and solvable through the application of science and technology."[39] Their concern is that students who learn about sustainability from only this perspective will not be attuned to the ethical and social aspects of sustainability.

The CCSSI and NGSS call for coherent learning progressions across the grades, emphasis on real-world learning, critical analysis based on evidence, problem solving, and integrative learning across disciplines. They thus represent movement toward an educational framework more consistent with the goals of ESD and other forms of sustainability education. These standards are not without controversy, however, and face a number of instructional and political challenges that according to some analysts raise important questions about their long-term prospects for success.[40] After widespread adoption by the majority of states, the Common Core standards have been the target of growing resistance from parents, teachers, and school officials, triggered in part by administration of the first Common Core tests in spring 2015 and a revolt against what is increasingly perceived as onerous and excessive use of standardized testing.[41] Part of the political uproar stems from assertions that, although the Common Core standards are not a national curriculum, they represent an unacceptable intrusion by the federal government into local control of education.[42] As of the time of writing, only thirteen states have formally adopted the NGSS. The standards' explicit reference to human-caused climate change has prompted several states to reject them, and lawsuits have been brought on these grounds in Wyoming, Michigan, and West Virginia.[43]

Educational systems have dynamic complexities of their own that make the implementation of sustainability education a challenge. The authors of the final report for the DESD note: "Education systems are complex and involve multiple levels of decision-making on educational policy and its implementation within schools, institutions of higher education, the workplace and communities. Advancing changes within education systems have required interventions at many different levels and have involved a broad cross-section of stakeholders."[44] The art of preserving opportunity across generations is inescapably political, and its legitimate and effective exercise should be first and foremost through truthful and reasoned instruction and persuasion. Public education in sustainability would be an essential aspect of this, but such education is likely to remain a marginal aspect of schooling until there is a sufficiently wide understanding of sustainability across the leadership of civic and social institutions.

Scholarly Criticisms of ESD

Scholarly assessments of the idea of ESD have been mixed. Sustainability has obvious significance for development—the path it should take and the extent to which it is possible—so the incorporation of sustainability content into DE should be uncontroversial.[45] It may indeed contribute to DE itself being more widely acceptable. What most concerns development educators is perhaps that the implementation of ESD has not proceeded much beyond preexisting environmental education and the emphasis on "green" schools.[46] The significance of development for environmental protection and sustainability is less obvious, and many environmental educators have perceived ESD as a threat to sound environmental education.[47] To the extent that Oxfam and other NGOs have filled the ESD implementation void with global citizenship instruction, it is unclear whether the understanding of environmental and sustainability problems has been advanced much.[48] Others who see the need for some educational response to problems of sustainability question whether ESD is a coherent and well-defined package.[49]

The various definitions of sustainable development are not all compatible, and none resolve basic tensions between sustainability and development.[50] What is entailed by "balancing" improvement in living standards with environmental stewardship? There are obvious tensions between development and sustainability, and there is no guidance in the DESD framework for how to make those tradeoffs—apart from intimations that demands on resources will need to be restrained in the affluent North. How much risk of ecological and societal collapse can be justified by the benefits of development? What exactly should be sustained? How is the role of *global justice* in ESD—its announced aim of instilling commitment to peace, justice, and human rights—to be understood? Is ESD simply a repository for diverse good causes, or is it an unavoidably interconnected whole? Are the causes that animate the ESD movement not as good as they appear, but actually in large measure those of the dominant neoliberal development paradigm?[51]

The main criticisms of ESD are as follows:

1. *ESD is incoherent.*
2. *ESD is too prescriptive.* It aims to inculcate values and shape behavior and thereby fails to respect liberty rights and other democratic values.[52] The word *for* in *education for sustainable development* and *education for sustainability* is problematic.[53]
3. *ESD undermines environmental education.* It is shaped by a corporate-friendly, neoliberal development paradigm, which is fundamentally incompatible with environmental protection. It makes sustainable development "the new aim of environmental education."[54]
4. *ESD does not rest on a sound conception of education.* In addition to being too prescriptive, its conception of education is instrumental and deterministic. It attempts to transmit knowledge and values in the service of social reproduction instead of allowing social construction of knowledge and values to occur in the service of socially transformative citizen engagement.[55]

Answering the Criticisms

ESD is incoherent. One version of this charge is that ESD is so ill-defined that it would be better to wait for a clearer roadmap of what it would be than to take steps toward implementing it. A fair response would be to reiterate UNESCO's view that local initiative and choice in adapting ESD to local circumstances are critical to success and local ownership of ESD. What is most critical to success is that

• teachers have adequate opportunities and incentives to learn what they need to know to provide their students with education that will be useful to them in meeting the challenges of sustainability; and
• required curricula and accountability schemes make room for the disciplinary foundations, cross-curricular linkages, and collaborative learning needed for students to understand and think critically about matters of sustainability and how they will live their lives.

The evidence suggests that few teachers have a deep enough understanding of sustainability—of the challenges, obstacles, ethics, and complexities outlined in chapters 1–5—to teach it well. Meaningful ESD implementation could focus on teacher preparation and opportunities to collaborate and does not need a better definition of SD or ESD to do so.

Unresolved tensions in the idea of *sustainable development* itself would remain, but educators and education policymakers will never be in a position to resolve those tensions themselves.[56] A good illustration of the tensions is the phenomenon of perilously unsustainable reliance on bushmeat in central Africa, where there is no obvious way to "balance" development and sustainability or to simultaneously "value and preserve" traditional cultures, "restore the state of our Planet," and ensure that "all people have sufficient food."[57] With human populations growing, hunting proceeding at a rate estimated at seven times what is sustainable, and cultures being dependent on hunting as they have been for eons because other forms of food production are not viable in that region, something has to give. Educators are in no position to prejudge how such dilemmas are best resolved, but sustainability education can nevertheless lay vital groundwork for *students* to productively examine the relationships between sustainability and development and discuss and investigate constructive responses to the problems of sustainability. The DESD framework seems to acknowledge this when it calls for developing the critical thinking and problem-solving skills learners will need to address "the dilemmas and challenges of sustainable development" confidently.[58] It would nevertheless make sense to decouple the conceptualization of sustainability education from the problematic idea of sustainable development and place sustainability front and center. To insist on some conceptual daylight between sustainability and development is not to expunge development from the content of an education in sustainability, however.

A second form of the criticism that ESD is incoherent is that SD is a collection of diverse good causes, not an indivisible whole. One answer to this is that development requires sustainability—in the long run, at any rate—and sustainability may in *some* ways be advanced by development. We noted in chapter 1 that the extent to which economic development is broadly conducive to sustainability has been greatly exaggerated, but we also have noted throughout this work that sustainability problems are largely problems of social coordination that can only be adequately managed through regional and global cooperation. It is a fair assumption that in many parts of the world the feasibility and legitimacy of such cooperation will require a strengthening of social capital along with water, food, and energy security sufficient to enable the great majority of people to adopt a long-term perspective on resource management.[59]

To defend the substance of UNESCO's vision of ESD, one need not identify in advance the terms of global justice or insist that future global citizens choose equitable international agreements as a rational response to problems of sustainability. It is enough to see that there is a prima facie case that it is rational to negotiate such agreements; success in addressing the problems is unlikely without global coordination, and equity is essential to reaching agreement on the terms of cooperation. These considerations do not provide conclusive grounds for regarding sustainable development—with its commitments to equity, mutual respect, and nonaggression—as an indivisible package. They do, however, provide sufficient grounds for regarding ESD as a coherent (if conceptually muddled) package. It is important that education prepare everyone for rational and informed participation in deliberations about the terms of possible global cooperation. The education in question should provide learners with a sound understanding of the history and prospects for international cooperation and engage them in critical scrutiny of the arguments for and against negotiating fair international agreements on greenhouse gas emissions and other aspects of sustainability. Thus understood, education in sustainability prepares learners for respectful, engaged global citizenship and openness to participating in fair terms of global cooperation, but it does not prescribe specific terms of cooperation.

ESD is too prescriptive. The first half of the answer to this criticism is implicit in our answer to the previous criticism: Sustainability education can and must be properly educative and not unduly prescriptive. The tensions between different values implicit in such ideas as sustainable development and adapting to preserve distinct cultures should not be prejudged by teachers. The virtues of self-restraint, justice, and respect for others are indisputably virtues, however, and it is not beyond the authority of teachers to nurture such virtues, so long as they understand that no disposition is a true virtue unless it is guided by the learner's own good judgment. As we explained in chapter 3, moral and civic virtues cannot be cultivated without cultivating virtues of intellect as well.[60]

ESD undermines environmental education. It is possible that many advocates of ESD understand sustainable development in neoliberal development terms, but no such understanding is evident in the DESD framework. The framework calls for providing the critical thinking ability needed to "transform current societies to more sustainable societies," and the nature and extent of that transformation is left open.[61] Therefore, there is no obvious basis for saying that UNESCO's ESD aims to transmit knowledge and values in the service of social reproduction rather than transformation. The environmental studies resources provided with the DESD framework are quite robust, in fact, and anyone who understands the conditions of sustainability outlined in this book will grasp that environmental science is a crucial aspect of ESD and that sustainability requires a serious rethinking of economic orthodoxies. The need to rethink economic orthodoxies is essentially acknowledged in ESD materials such as the ESD Toolkit 2.0, and it is acknowledged by development economists who adopt a human development perspective on international development.[62]

ESD does not rest on a sound conception of education. What lends this criticism plausibility is that advocates of ESD and EFS have not devoted much effort to grounding their conceptions of sustainability education in a guiding philosophy of education. We outlined a broad philosophy of education in chapter 4 and will now show that ESD or EFS as we conceive it (EiS) is not just compatible with but also essential to a sound education. Environmentalists and environmental educators have been suspected of having a radical, nature-focused agenda that does not serve human interests or the interests of students but instead trains students to act *for* the environment or *for* sustainability. It should be clear by now that human interests are critically dependent on the health of natural systems, but in conceptualizing education in sustainability it is nevertheless important to begin from a conception of education that predicates educational requirements on what is good for students and equips them to live well.

We suggested at the opening of this chapter that children are entitled to education in sustainability and that there are reasons of both morality and prudence to provide an education in EiS that will facilitate cooperation to solve the problems of sustainability. We will expand on these arguments that EiS is not just educationally legitimate but essential to an adequate education, building on the general conception of education introduced in chapter 4.

Why All Children Are Entitled to EiS

The opportunity to acquire the understanding, capabilities, and virtues that equip one to survive and live well in the world one will inhabit arguably is a basic entitlement. We have argued that the institutions of society should provide the necessities for living well that cannot be efficiently procured by individuals without aid, and failure to do so within the society's sustainable capacity would constitute injustice. Children are *entitled* to substantial opportunities to live well, and educational institutions are clearly among those critical to providing such opportunities, their role being to promote forms of personal development essential and conducive to living well.

The challenges and obstacles to sustainability outlined in chapters 1 and 2 are fundamentally important aspects of our world that will shape opportunities for the foreseeable future. How could an education suitable to providing substantial opportunities to live well not provide the understanding and ability to contend with these challenges and obstacles? Because obstacles to sustainability include collective choice problems that cannot be solved without coordination—without multiscale cooperation, including *global* cooperation—how could the education children are entitled to not include preparation to understand and participate in regional and global cooperation? How could it not include education for rational and informed global citizenship?

Educational institutions are distinguished by the fact that they promote these forms of development by initiating learners into practices that *express human flourishing*—practices through which they can fulfill human potentials in admirable and satisfying activities of a good life and satisfy basic psychological needs for competence, autonomy, and mutually affirming relationships. Many such practices contribute to civic competence and economic and social opportunity, and practices of inquiry and critical rationality or critical thinking have further significance as tools of personal efficacy or effective agency. There are three basic aspects of *agency*, or being an actor in the world—the beliefs or understanding of the world on which we rely, the goals and values we act from, and the capabilities we exercise in acting—and these are all aspects of ourselves for which we can and should take responsibility, inasmuch as we can engage in more or less effective self-reflection on all of these aspects of ourselves and strive to be better. Engaging in such self-examination allows us to selectively overcome the limitations and self-defeating aspects of our thought patterns, understanding, abilities, motivation, and preferences. It makes us freer by degrees and more effective in our efforts to live well. Providing education in the practices of critical thinking can thus be defended as an important aspect of the social and educational enterprise of providing children with substantial opportunities to live well.

An important related goal of education is to cultivate good judgment, which is an aspect of competence in everything a person does. We act in contexts embedded within other contexts—social, institutional, built, and natural—all dynamic, interacting, and changing, and the ramifications of our daily acts reach every corner of the globe in ways that are endlessly complex and spill across a growing range of increasingly specialized forms of expertise. Understanding the significance of our choices would seem to require either a command of and ability to connect all of these forms of expertise or informed reliance on the judgments of diverse experts and a remarkable ability to bring those judgments to bear on one's own ideas and choices. The former is impossible and the latter far from easy, but understanding basic aspects of how serious inquiry works and how complex systems function makes it easier to discern who the relevant experts are and what their findings imply for one's interests and life projects.

A school designed to cultivate good judgment would both enable students to make the perspectives of individual disciplines their own and foster skill in cross-curricular integration of those distinct perspectives.[63] Breadth of experience, reading, and human contact are essential resources for critical thinking, and it is not hard to imagine promoting such breadth through forms of experiential and project-based team learning that foster cross-curricular integration and connect the classroom to the community and the world. Robust critical thinking also deploys a variety of analytical tools and is guided by a variety of standards of judgment, some general (those of general-purpose logic, argument analysis, and the like) and some specific to the forms of evidence and reasoning that have been developed in diverse domains of inquiry and practice.[64] These must be learned, usually with the help of teachers, and the art of bringing diverse analytical tools and standards together in understanding and judging complex matters must be practiced. The most robust lessons in critical thinking and judgment might allow students, working in teams, to bring the intellectual resources of diverse fields of study to bear on real-world problems, research those problems, and develop proposals.

If this is what education is, then EiS would not be education if it were indoctrinating, narrowly behavioral, inherently predicated on disputable value commitments, or instrumental in a way that gave priority to ends external to the development and well-being of learners themselves. Children are entitled to education and to a form of EiS that provides them not just the breadth and depth of understanding and skills necessary for coping in a perilous world but also the critical insights and habits of mind needed for self-direction in overcoming the obstacles to sustainability, and the creative and collaborative capacities needed not simply to meet the challenges of sustainability but also to find satisfaction in so doing. Appropriate and effective EiS would be socially transformative in ways that would improve the prospects for human well-being.

Reasons of Morality and Prudence for Providing EiS

Our individual actions have a global reach and impact through economic relations, pollution, reliance on common resources, and climate destabilization, and we argued in chapter 3 that this makes the negotiation of global regulatory agreements a moral and not just practical necessity.[65] *Morally*, we owe it to each other to discuss the ways our actions impair each other's interests and to settle what will and will not be recognized as wrongful violations of those interests. *Practically*, there are no unilateral solutions to global pressures on planetary boundaries and unsustainable transboundary burdens on water systems and other forms of natural capital. Coordinated action to address these problems is needed, often on a global scale, and it is inconceivable that such action will occur except on the basis of agreements that are perceived as fair.

If global cooperation is a requirement of both morality and prudence, then education for global citizenship is required on two grounds. It is required first of all because the *facilitation* of cooperation requires

• well-informed understanding of the value of cooperating;
• understanding of the possible institutional bases of cooperation, such as the United Nations; and
• the understanding, skills, abilities, and knowledge that may be needed to participate in cooperative arrangements.

A second related consideration is that education for global citizenship is required to secure the *legitimacy* of whatever terms of cooperation might be negotiated. Legitimacy rests on transparency, which requires understanding of what is at stake, hence a wealth of relevant education for all who may directly or indirectly be parties to the negotiation or subject to the terms of cooperation it yields—in short, everyone in the world. Legitimacy also requires that the terms of cooperation be imposed (that compliance be obtained) as much as possible through voluntary cooperation based on a sound understanding of what is at stake and how the cooperative arrangements have been arrived at. However far we may be from settling on fair terms of global cooperation, it is not too early to lay the groundwork for this to occur in a way that respects the ideal of full transactional transparency.

As we have said, global cooperation to address problems of sustainability is not only morally required, but also *prudent*. Faced with the declining health of global commons—the atmospheric and oceanic systems—on which human civilization depends, the longer the nations of the world approach these matters competitively and without self-imposed collective limits, the more difficult life will become. This makes it not just morally compulsory but prudent to favor the cooperation for which we have just outlined the educational requirements. As a matter of prudent self-interest, informed and rational readiness to cooperate is a dimension of the education children are entitled to as one aspect of educational preparation to live well.

To the extent that this concerns climate disruption, a contrary view that retains a grain of truth is that there will be "winners" and "losers." What is true is that *below some threshold* of warming, some people are more vulnerable than others.[66] It does not follow from this that it would be prudent of those who may be currently less vulnerable to withhold the cooperation necessary to stay within that threshold. In the absence of effective coordinated action, there is little assurance there will be any winners much longer, and even less reason to think the interests of the winners of today wouldn't be harmed enough to make cooperation worthwhile. A planet altered to the profound detriment of most forms of life on it for hundreds of years, if not far longer, is not in the interest of anyone who feels any personal stake in a civilized future for humanity.

There are, to be sure, countless unresolved questions about the terms of the cooperation needed. Most vexing may be the tradeoffs between population size and quality of life. Faced with a status quo that allocates substantial reproductive rights through markets that will increasingly deny those who are poor the means to sustain the lives of any children they bear, one can speculate that it may before long seem fair and most humane to conceive of reproductive rights not as unlimited liberty rights, but as limited welfare rights. What can be said here and now is that the various reasons adduced in support of the education we are defending are all reasons to believe that everyone in the world should be enabled to make reproductive decisions informed by knowledge of the momentum and impact of unsustainable human demands on the planet. Universal basic education for girls would be a critical first step.

We have argued that children are entitled to a form of EiS that is properly educative and that part of such EiS is preparation to prudently participate in global cooperation to address problems of sustainability. We have also argued that there are reasons of both morality and prudence to be ready to participate in fair terms of global cooperation and that such cooperation must rest on a rational, free, and informed understanding and acceptance of the terms of cooperation. The prospects of reaching agreement on any such terms and of securing compliance with them rest similarly on understanding and acceptance of this kind—and hence on the quality and global reach of EiS.

Proposals for Curriculum and Instruction

The recommendations to which all of the foregoing considerations point can now be stated.

Respect children's right to know, think for themselves, and use their own good judgment in living responsibly and well. Respect teachers' professional judgment and provide them the opportunity to learn what they will need to know and acknowledgment of success in providing innovative instruction. The fundamental task of education is to promote forms of development conducive to living well, and in doing so provide substantial opportunities to survive and flourish. The development of self-reflection, critical thinking, and capacity to live creatively is basic to this and to the parts of education specifically related to enabling children to meet the challenges of sustainability. To deliver a suitably educative form of EiS, teachers will need adequate opportunities and incentives to acquire the depth of understanding of sustainability they will require, and curriculum and accountability frameworks will need to make room for cross-curricular collaboration and innovative instruction.

Teach environmental studies more systematically. The environmental studies curriculum should include the relevant science, the problems, the regional and global distribution of impacts, and the state of cooperation or noncooperation in solving the problems. It is important that students not just learn science, but be able to understand what is happening in the world around them through the methods and explanatory resources of science. Without such understanding and openness to scientific inquiry, science learning will be inert and offer little foundation for conviction about what must be done. To say that adequate education in environmental science should be supplemented with other forms of sustainability education— concerning the problems, distributions of impacts, and state of cooperation or non-cooperation—implies an extension of student learning and inquiry into matters that are vitally important, but also controversial. Students in affluent countries will need to understand far more than most now do about poverty and the distribution of environmental benefits and burdens both globally and within their own countries. To understand these things, they will need to learn some distressing truths about the vast resources devoted to suppressing the truth about pollution and environmental damage and their enormous and inequitably distributed burdens on human health.

Integrate environmental studies with honest history and prehistory. An understanding of sustainability requires knowledge of the patterns and dynamics of societal complexity, collapse, and survival. Students should all know the story of Easter Island, of irrigation and ecological catastrophe in the Fertile Crescent, and the population overshoot and collapse of the Maya. They should understand the analytical models through which anthropologists, geographers, and others understand societal complexity and collapse, and they should understand how those models apply to their own world and lives.

In the spirit of UNESCO's wide conception of environmental literacy, history and social studies curricula should also address the role of competition for scarce resources in the genesis of war and genocide.[67] This is best accomplished through detailed case studies, including some that provide background for ongoing conflict. It is admittedly not easy in schools to address honestly such controversial matters as the role of water rights in the Israeli-Palestinian conflict or US and British oil interests in the political history of the Middle East. It is nevertheless very much within the rights and in the interest of students in the United States, United Kingdom, and Middle East to know the specific truths and general lessons of such history. It is surely in their interest to understand the extent to which the case for war is often built on obfuscation, disguising unacknowledged private interest as a common national or international interest. In the age of scarcity humanity is facing, global citizenship and security will require vigilant resistance to the seductions of war.

Integrate economics with these environmental studies. Students can benefit from honest instruction in production methods, both agricultural and industrial, and the environmental controversies surrounding them. They should come away knowing what amounts of energy and water resources are consumed in the production of common products (eight hundred gallons of water for a hamburger, for instance, and 1,700 gallons of water for a gallon of corn ethanol) and how atmospheric carbon and other wastes are released. Here and throughout the curriculum, the focus on critical and inventive thinking should be as effective as possible and engaged with vital questions: How can production, marketing, and distribution systems be redesigned to be more environmentally friendly? How can we best live without economic growth, if growth is precluded? To what extent would more egalitarian distributive policies help? Who, if anyone, gains from growing populations and growing demand for property, goods and services, and who does not?

Encourage resourcefulness, inventiveness, and adaptability. Schools can promote a readiness to examine, rethink, and redesign every aspect of how we live in order to promote the most optimal adjustment to a sustainable human footprint. In connection with such promotion, they can nurture skills in diverse practical arts, with an emphasis on design, economy, and adaptability. This approach may be considered not only collectively beneficial as a basis for adapting how we live but also a form of individual insurance against occupational risk in a world of rapid change. It would facilitate the self-provisioning that Juliet Schor has identified as an important step individuals can take toward a more sustainable and humane economy.[68]

With regard to adaptability and the potentially calamitous price of cultural intransigence, schools should provide lessons in the hardship that societies have brought upon themselves by failing to adapt. The failure of the Greenland Norse to abandon maladaptive aspects of their European heritage provides one vivid illustration of such a hardship.[69] Schools could provide case-based investigative learning opportunities focused on contemporary cultural practices, such as those pertaining to gold. In the interest of dietary adaptability, which could reduce the carbon footprint in the affluent North by 20 percent, schools could provide students with instruction in the energy economics of food choices and experiences designed to broaden their culinary horizons.

Encourage the enjoyment of low-impact activities as a basis for living both well and in ways compatible with sustainability. Environmentally low-impact activities could include intellectual, musical, athletic, and social pastimes, with modifications from their present forms as needed, such as to reduce demands on transportation and energy resources. The growing body of research on well-being and materialism that we discussed in chapter 3 can be taught in connection with this and as one disciplinary foundation for critical thinking about cultural practices.

Decommercialize schools. Commercial messages to consume, define one's identity through consumption, and address one's problems though material consumption should be banned from schools except as objects of critical thinking exercises. This would be a useful if modest step toward making it easier for children to distinguish between what they need and what they want and for them to resist inducements to excessive and imprudent consumption. Modest as this step would be, it requires that we muster renewed belief in the importance of preserving public spaces, including schools, in which we can conduct the public's work, in the public interest, through public reason, and free of the dominating influence of commercial interests.

Teach critical thinking and enable children to distinguish the truth from propaganda. It bears repeating that throughout the curriculum, the focus on critical and inventive thinking should be as effective as possible and engaged with vital questions. This requires sustained, direct instruction in methods of critical thinking that draws on multiple disciplinary perspectives, practice in thinking critically and creatively about the vital questions at issue, and a critical study of media and propaganda. We owe children the where-withal to protect themselves against campaigns of misinformation and misleading argument.

Encourage critical self-reflection and creative living through literature and the arts. Stories communicate important human truths in vivid and memorable ways. The sense of history that may guide us in living well is most easily brought to bear through narrative patterns we can recognize in our own lives. In this regard, such patterns are similar to analytical models of the dynamics of ecosystems and social complexity and collapse in supplying tools for analysis and critical thinking. Beyond this, the arts in general can contribute vitally to nurturing an ability to think beyond the norms and conventions of daily life.

Use collaborative, civic, and project-based learning. Learning how to think effectively about sustainability and how to cooperate to achieve it requires highly engaged cross-curricular learning and the experience of working with others toward common goals. It is advanced by experience in engaging others unlike oneself in respectful and productive discussion and collaboration. Collaborative, civic, and project-based learning have obvious value in this domain, as in others. Public health risks in the community can be a particularly engaging focus for project-based learning, with obvious links to environmental concerns, critical thinking about regulatory schemes, and health, home, and career education.[70]

Prepare children for global cooperation. There is much that could contribute to an openness and ability to participate in global cooperation, beginning with serious instruction in geography, languages, world affairs, the history of the United Nations, an understanding of poverty and the faltering capacity of governments to secure a livable future for their citizens, and collaborative citizenship learning. If patriotism is to be encouraged in schools, it must be a form of patriotism that is devoted to protecting what is good in and beyond one's country and correcting what is not good—a form of patriotism that is consistent with responsible global citizenship.[71] Schools should facilitate global citizenship through student involvement in global constitutional activity and nurture sympathetic attachment to a global civic community.

Prepare everyone for a world with lower fertility rates and the prospect of fewer human beings. If the human presence on Earth is as environmentally over-extended as current estimates suggest and the UN's population projections are correct, then the human population of Earth will decline through the latter half of this century. What is less certain is how humane or inhumane the path of descent will be and how and to what extent decisions will reflect the trade-offs between population size and risk. It is not clear how best to prepare young people for this, but any approach taken must seek to equip them with the imagination, understanding, and critical capacity to make rewarding lives for themselves in ways we may not now envision.

Ethical Components of EiS

The ethical components of EiS will no doubt be challenged as tantamount to indoctrination, teaching a controversial or false morality, or exceeding a public school's authority to educate simply because the subject matter is ethical. We laid some groundwork for defending a role for sustainability ethics in schools in our justifications of the ethic of sustainability itself in chapter 3. We need to say more, however, especially regarding the manner of instruction, which should be to engage students in ethical inquiry. It is important to recognize at the outset that legitimacy must be understood not just externally in the public domain in which the work of schools is questioned and challenged by adults, but also internally in the life of schools and classrooms. Legitimacy concerns the justification of an exercise of authority to those subject to the authority, and the success of educators' efforts to provide students with opportunities through learning requires informed cooperation, much as the political art of preserving opportunity for the long-term requires it.

The principles of sustainability ethics identified in chapter 3 are applications of familiar and indisputable principles of common morality, formulated to make their application to matters of sustainability evident. The ways in which they reflect basic respect for fellow human beings should be evident to anyone who acknowledges there is reason to be concerned about sustainability. Therefore, the principles of sustainability ethics we have identified should be teachable in schools as no less self-evident than other aspects or applications of common morality. Skeptics about morality may insist that even the most basic provisions of an ethic of respect for persons are mere "opinion" and not legitimately teachable in public schools, but all common law jurisdictions *enforce* this ethic through legal penalties, and it cannot possibly be legitimate to do so and yet fail to *teach* it.[72]

We argued in chapters 2 and 3 that the legitimacy of law presupposes a foundation of education in the moral content or justification of laws. It follows from this that public schools have not only the authority to engage in basic moral education but also a responsibility to do so. Our view of moral education and children's progress toward moral maturity is that it begins in basic moral principles and examples of goodness and badness and progresses by adding an increasingly nuanced understanding of complexities that arise in applying those principles and living a good life.[73] The progress from an immature grasp of human affairs and correspondingly inadequate understanding of the reach and limitations of common moral principles to more mature and adequate understanding is slow, is sometimes painful, and generally requires reflection on concrete cases. Those cases may be real or fictional. However the cases are encountered, the learning mediated by reflection or examination of the cases will be modeled on the learner's experience of and participation in morally serious *conversation*—conversation that seeks understanding and is thus open-ended and to some extent philosophical. Students are not only capable of such case-based conversation but eagerly embrace it.[74]

The teachers who can succeed in this enterprise will be ones who are themselves practiced in norms and patterns of ethical reflection and have earned moral authority with their students, in part through exemplifying the requisite virtues, both reflective and interpersonal. Engagement in ethical reflection must be made attractive to students in the way it is modeled, through the benefits it promises, and through the immediate rewards it brings in satisfying students' needs for competence, self-determination, and positive relationships.[75] These benefits and rewards are possible only if teachers communicate the value of moral seriousness and lead students in activities of moral reflection. The norms of reasonableness and rational exchange must be made explicit and reinforced to create a classroom community of cooperative ethical inquiry. The participation to which students are invited must be predicated on acceptance of the responsibility to respect others through attentive openness to learning and to being persuaded by them, as well as on a corresponding right to establish moral authority of their own through the ethical perceptiveness of their remarks. The foundations of a teacher's moral authority are undoubtedly wider than the terms of her engagement in ethical instruction, but the quality of this engagement will surely matter.

A teacher needs not merely to be *in authority*—in a position of authority in her classroom, in the sense of having an institutionally conferred right to teach and manage her class—but to *have authority* with students, in the sense of being able to procure their cooperation through their belief that she knows what's best—their belief that she "is in a position to know [what] is to be done."[76] Such authority will be perceived by students as noncoercive to the extent that they perceive her as having *moral authority*. Furthermore, they are likely to perceive her as having such authority—as knowing what is for the best all things considered—if their experience of her is that she respects them, cares about them, and manages her classroom in a way that aims effectively at the good of all her students. The efficacy of teaching and exercises of classroom authority depend upon the teacher's success in establishing moral authority, and this applies as much to teaching ethics as to teaching anything else. Engaging students in open-ended ethical reflection can also enhance a teacher's moral authority through the respect it shows students and through its potential to enable students to better understand what is in their interest and how it guides their teacher's actions. Precollegiate ethical instruction through philosophical inquiry may thus function much as it would in the city of Plato's *Laws*: as both a form of instruction fundamental to promoting a person's good, a good that involves guiding oneself by one's own reason, and as a form of instruction that contributes to the legitimacy and rational acceptance of reasonable authority by enabling a person to better see what is and is not conducive to her own well-being.

It has been shown by others in compelling detail that what teachers can do is limited and shaped by the character and mission of the encompassing school.[77] This being so, we must—even as we promote the classroom as a community of ethical inquiry—look beyond the classroom to the mission and moral tenor of schools as a whole. What is needed in schools is compatible with the picture we have sketched of what is required of individual teachers and conducive to establishing noncoercive authority relations. As educational psychologist Joan Goodman writes: "The mission of academic excellence can take on a more moral and collective texture when excellence is extended from self-serving attainments to *valuing deep exploration and articulation of issues*, high standards in a range of endeavors, and personal attainments oriented to improvements outside the school doors. These grander more moral objectives, I suggest, are platforms that better legitimate authority and its distribution."[78] It is hard to think of anything humanity needs more at this time than "valuing deep exploration and articulation of issues" pertaining to sustainability. And—as distressing as it may be to many adults—making such exploration and articulation a focus of our schools would be a powerful way to communicate our concern for the well-being of our children and enlist them as partners in living well in a world of opportunities shaped and threatened by the way we have been living. Initiating them into living well in ways that do not destroy the opportunities of others to live well is the only basis on which we can legitimately establish the moral authority to educate at all.

Recalling the costs of market credentialism identified in chapter 4, we must note that providing rising generations with opportunities for shorter and more certain pathways to occupational certification might give them the peace of mind to put their sustainability-related learning to work in preserving opportunities for others to live well in the future. In the context of a larger dynamic of growing complexity and stratification of opportunity, problems of motivation and the legitimacy of educational authority are matters of fundamental and urgent concern for the education of children who are not privileged and have little hope of outcompeting others in a lengthening and costly quest for diplomas and degrees. Reflecting on what we can offer students whose prospects are already bleak also offers insight into what we should aim to give everyone in the years ahead. Anxiety and despair are common responses to learning about sustainability, and the best antidote is a sense of personal efficacy and well-being grounded in understanding, action, belonging, and the security of having opportunities one can pursue. It will help a great deal if we focus on these essentials of personal well-being and motivation, as well as on the larger context in which education in sustainability occurs.

Conclusion: The Promise of Opportunity

Human parents have not only summoned their children into life through conception and birth, they have simultaneously introduced them into a world. In education they assume responsibility for both, for the life and development of the child and for the continuance of the world. ... the child requires special protection and care so that nothing destructive may happen to him from the world. But the world, too, needs protection to keep it from being overrun and destroyed.
—Hannah Arendt[1]

We remarked in the preface to this book that in parenting and teaching we face the awkward reality of being spokespersons for the world—adults who must believe in the prospects for living well on this planet and in our capacity to equip our children and students to live well without destroying those prospects for others. We must believe in a world of opportunity but must also recognize that we are collectively living in ways that are diminishing the opportunities in which we must believe. It would not be unfair to observe that to persist in this condition is a collective act of bad faith, a betrayal not just of an abstract social contract but of the young we cherish.

Making good on this promise of opportunity will require that we educate children well and reform the institutions, systems, structures, settings, policies, and practices that structure in many ways and to varying degrees the activities of contemporary life that have overshot the limits of what nature can bear. We have argued through the course of this book that whatever technical innovations may be required to sustain civilization as we know it, sustainability is not a science but an art of social coordination. It is in this sense an art of governance—and one that can be effective and legitimate only if it creates the conditions for widely distributed cooperation based on common understandings and sufficient opportunities for everyone to live well now. We saw this in Elinor Ostrom's work on the strengths and requirements of distributed self-governance of environmental commons, in reflections on the coordination costs and hazards of complexity, and in the literature on the *wickedness* of sustainability problems—the term *wickedness* reflecting, somewhat perversely, a dawning realization that attempts to command and control people are only marginally useful in improving how social systems function and interact with natural systems. We have argued in this book that if contemporary societies are only now coming to grips with these matters, there may be benefit in revisiting the ethical and political thought of Plato and Aristotle, who, for all their limitations, understood that moral clarity, education, and fair opportunity are fundamental to governance, whereas law—important as it is—plays a secondary role and is most effective when it is understood as not so much coercive as educative. All of this implies that government regulation, collective action guided by common norms and understandings, and market mechanisms are less distinct than imagined and must be harmonized to protect opportunity both now and in the future.

The educational reforms we call for are a foundational aspect of the larger package of reforms that are needed. They are foundational in the sense that they would provide a foundation of understanding, competence, and civic virtues helpful to enacting and implementing reforms in all civic spheres sufficient to putting humanity on a sustainable trajectory. Although we have affirmed the desirability of a variety of noneducational reforms in the course of this book, our purpose has been much more to provide general normative and conceptual guidance than to offer a systematic agenda for reform. Our understanding of living well inspired by Aristotelian ideas and recent developments in eudaimonistic psychology offers a basis for living more sustainably and for conceptualizing what it would mean to preserve opportunities to live well into the future. It also offers a basis for fulfilling our responsibilities to both children and the world.

Notes

Preface

1. Randall Curren, *Education for Sustainable Development: A Philosophical Assessment* (London: PESGB, 2009).

Introduction

1. See, e.g., David Archer, *The Long Thaw: How Humans Are Changing the Next 100,000 Years of Earth's Climate* (Princeton, NJ: Princeton University Press, 2009); Justin Gillis, "U.S. Climate Has Already Changed, Study Finds, Citing Heat and Floods," *New York Times* (May 7, 2014); Michael J. Graetz, *The End of Energy: The Unmaking of America's Environment, Security, and Independence* (Cambridge, MA: MIT Press, 2013); Wil S. Hylton, "Broken Heartland: The Looming Collapse of Agriculture on the Great Plains," *Harper's Magazine* 325, no. 1946 (July 2012); Bill McKibben, "The Pope and the Planet," *New York Review of Books* 62, no. 13 (August 13, 2015); Suzanne Moser and Maxwell T. Boykoff, eds., *Successful Adaptation to Climate Change* (London: Routledge, 2013); Adam Nagourney, Jack Healy, and Nelson D. Schwertz, "California Image vs. Dry Reality," *New York Times*, April 5, 2015; Jeffrey Sachs, *The Age of Sustainable Development* (New York: Columbia University Press, 2015); and Gernot Wagner and Martin L. Weitzman, *Climate Shock: The Economic Consequences of a Hotter Planet* (Princeton, NJ: Princeton University Press, 2015).

2. See, e.g., Thaddeus R. Miller, *Reconstructing Sustainability Science* (London: Routledge, 2014); Bert J. M. de Vries, *Sustainability Science* (Cambridge: Cambridge University Press, 2013); Hiroshi Komiyama and Kazuhiko Takeuchi, "Sustainability Science: Building a New Discipline," *Sustainability Science* 1, no. 1 (October 2006), http://www.springerlink.com/content/214j253h82xh7342/fulltext.html; and Eli Lazarus, "Tracked Changes," *Nature* 529 (January 2016).

3. This is essentially the approach of Kent Portney, *Sustainability* (Cambridge, MA: MIT Press, 2015), chap. 1, "The Concepts of Sustainability."

4. See Bernard Gert, *Common Morality: Deciding What to Do* (New York: Oxford University Press, 2007).

5. It is not uncommon for elementary ethics textbooks to begin by examining and setting aside moral relativism and subjectivism. See James Rachels and Stuart Rachels, *The Elements of Moral Philosophy*, 5th ed. (Boston: McGraw-Hill, 2007); Russ Shafer-Landau, *The Fundamentals of Ethics* (New York: Oxford University Press, 2010); and Russ Shafer-Landau, *Whatever Happened to Good and Evil?* (New York: Oxford University Press, 2004).

6. See, e.g., James Gustave Speth, *The Bridge at the Edge of the World: Capitalism, the Environment, and Crossing from Crisis to Sustainability* (New Haven, CT: Yale University Press, 2008), 126–146, 156–159; and Juliet Schor, *Plenitude: The New Economics of True Wealth* (New York: Penguin, 2010), esp. 176–180. The notable exception is Tim Kasser, who is both engaged in the fundamental psychological research cited by Speth, Schor, and others and a writer on sustainability himself. See Tim Kasser, *The High Price of Materialism* (Cambridge, MA: MIT Press, 2002); and Tim Kasser and Allen Kanner, *Psychology and Consumer Culture: The Struggle for a Good Life in a Materialistic World*, 2nd ed. (Washington, DC: American Psychological Association, 2013).

1   What Sustainability Is and Why It Matters

1. See Brian Barry, "Sustainability and Intergenerational Justice," in *Environmental Ethics*, ed. Andrew Light and Holmes Rolston III (Malden, MA: Blackwell, 2003). Barry writes: "The core concept of sustainability is ... that there is some X whose value should be maintained, in as far as it lies within our power to do so, into the indefinite future" (491), the forms of value in question being framed in terms of needs, opportunities, quality of life, freedom and democracy, and the intrinsic value of the natural world. Our view is that these framings of the value at stake all revolve around opportunity to live well, including the reference to the intrinsic value of the natural world, which might concern opportunities for sentient nonhumans to live well, the noninstrumental valuing of nature as an aspect of human beings living well, or both. See William FitzPatrick, "Valuing Nature Non-Instrumentally," *Journal of Value Inquiry* 38 (2004). For discussion of Barry's understanding of the preservation of opportunity to live well as neutral with regard to conceptions of living well and the alternative approach of conceiving (the nature or measure) of what should be preserved as natural capital or enumerated aspects of nature or natural items, see Andrew Dobson, *Citizenship and the Environment* (Oxford: Oxford University Press, 2003), 147–148 and 155–173; Bryan Norton, *Sustainability: A Philosophy of Adaptive Ecosystem Management* (Chicago: Chicago University Press, 2005), 119ff.; Alan Holland, "Sustainability: Should We Start from Here?," in *Fairness and Futurity: Essays on Environmental Sustainability and Social Justice*, ed. Andrew Dobson (Oxford: Oxford University Press, 1999).

2. A notable manifestation of this shift was the adoption of the language of sustainability by the World Conservation Union between 1980 and 1991. For a full history of the term *sustainability*, see Jeremy L. Caradonna, *Sustainability: A History* (Oxford: Oxford University Press, 2014).

3. See Thomas Wiedmann and John Barrett, "A Review of the Ecological Footprint Indicator—Perceptions and Methods," *Sustainability* 2 (2010); L. Blomqvist et al., "Does the Shoe Fit? Real versus Imagined Ecological Footprints," *PLoS Biology* 11, no. 119 (2013), doi:10.1371/journal.pbio.1001700; William E. Rees and Mathias Wackernagel, "The Shoe Fits, but the Footprint Is Larger than Earth," *PLoS Biology* 11, no. 11 (2013), doi:10.1371/journal.pbio.1001701; Jeroen C. van den Bergh and Fabio Grazi, "Reply to the First Systematic Response by the Global Footprint Network to Criticism: A Real Debate Finally?," *Ecological Indicators* 58 (2015). For resources related to footprint analysis, see the website of the Global Footprint Network: http://www.footprintnetwork.org.

4. Robert Costanza and Herman Daly, "Natural Capital and Sustainable Development," *Conservation Biology* 6, no. 1 (1992): 38.

5. Ibid.

6. Mathis Wackernagel and William Rees, *Our Ecological Footprint* (Gabriola Island: New Society, 1996), 35. See also Nicky Chambers, Craig Simmons, and Mathias Wackernagel, *Sharing Nature's Interest: Ecological Footprints as an Indicator of Sustainability* (London: Earthscan, 2000). The flow and recharge rates for the Ogallala Aquifer stretching under 134,000 square miles of the Great Plains of the United States are so low that much of the water in it is believed to be *paleowater* ("fossil water") dating to the last ice age or earlier. Apart from natural discharge into springs, streams, and the Platte River in Nebraska, it may be considered largely a passive reservoir that produced no "services" until extraction through wells became significant in the 1930s. It now supplies about 30 percent of all irrigation water in the United States and has in some regions (notably Texas, New Mexico, and Kansas) been drawn down five to six feet annually, or about one hundred feet since the early 1990s. See Edwin D. Gutentag et al., *Geohydrology of the High Plains Aquifer in Parts of Colorado, Kansas, Nebraska, New Mexico, Oklahoma, South Dakota, Texas, and Wyoming*, US Geological Survey Professional Paper 1400-B (Washington, DC: US Department of the Interior, 1984), http://pubs.usgs.gov/pp/1400b/report.pdf; and Wil S. Hylton, "Broken Heartland: The Looming Collapse of Agriculture on the Great Plains," *Harper's Magazine* 325, no. 1946 (July 2012).

7. Costanza and Daly, "Natural Capital and Sustainable Development," 39, 44. On page 39, the authors say that "constancy of total natural capital (TNC) is the key idea in sustainability of development," but the concepts of environmental accounting introduced to that point in their paper concern sustainability and do not depend on any conceptual link between sustainability and development.

8. Wackernagel and Rees, *Our Ecological Footprint*, 3.

9. World Wildlife Fund, *Living Planet Report 2014: Species and Spaces, People and Places* (Gland, Switzerland: WWF International, 2014), 9ff., http://wwf.panda.org/about_our_earth/all_publications/living_planet_report/.

10. Ibid., 33.

11. See UN Foundation, *The Millennium Ecosystem Assessment* (Geneva: UN Foundation, 2005), http://millenniumassessment.org/en/index.html. This was a comprehensive assessment coauthored by 1,350 scientists from ninety-five countries and twenty-two national academies of science.

12. Johan Rockström et al., "A Safe Operating Space for Humanity," *Nature* 461, no. 24 (September 2009): 472, http://www.nature.com/nature/journal/v461/n7263/full/461472a.html. Commentaries are available at http://tinyurl.com/planetboundaries. See also Worldwatch Institute, *State of the World 2013: Is Sustainability Still Possible?*, chaps. 2 and 3 (Washington, DC: Island Press, 2013); World Wildlife Fund, *Living Planet Report 2014*, chap. 2.

13. Rockström, "A Safe Operating Space for Humanity," 474. On the idea that Earth has entered a new, human-caused geologic epoch, the Anthropocene, see David Biello, *The Unnatural World: The Race to Remake Civilization in Earth's Newest Age* (New York: Scribner, 2016).

14. Will Steffen et al., "Planetary Boundaries: Guiding Human Development on a Changing Planet," *Science* 347, no. 6223 (2015), doi:10.1126/science.1259855. The authors portray the updates they offer as "one step on a longer-term evolution of scientific knowledge to inform and support global sustainability goals and pathways" (8).

15. See USDA Forest Service, Pacific Northwest Research Station, "The Healing Effects of Forests." *Science Daily*, July 26, 2010, http://www.sciencedaily.com/releases/2010/07/100723161221.htm; Kathleen D. Moore and Michael P. Nelson, eds., *Moral Ground: Ethical Action for a Planet in Peril* (San Antonio, TX: Trinity University Press, 2010). On environmental "cultural services," see World Wildlife Fund, *Living Planet Report 2012: Biodiversity, Biocapacity and Better Choices* (Gland, Switzerland: WWF International, 2012), 85–86, http://worldwildlife.org/publications/living-planet-report-2012-biodiversity-biocapacity-and-better-choices.

16. Experiencing one's life as meaningful is an aspect of living well, and it seems to require devotion to things one perceives and treats as independently valuable. See Susan Wolf, *Meaning in Life and Why It Matters* (Princeton, NJ: Princeton University Press, 2010); and Randall Curren, "Meaning, Motivation, and the Good," *Professorial Inaugural Lecture*, Royal Institute of Philosophy, London, January 24, 2014, http://www.youtube.com/watch?v=rhjZvbvpJYQ&feature=youtu.be.

17. Thomas Dietz, "Informing Sustainability Science through Advances in Environmental Decision Making and Other Areas of Science," paper presented at the NRC Sustainability Science Roundtable, Irvine, CA, January 14–15, 2016, http://sites.nationalacademies.org/cs/groups/pgasite/documents/webpage/pga_170344.pdf.

18. See P. A. Frugoli et al., "Can Measures of Well-Being and Progress Help Societies to Achieve Sustainable Development?," *Journal of Cleaner Production* 90 (2015); and Megan F. King, Vivian F. Renó, and Evelyn M. L. M. Novo, "The Concept, Dimensions and Methods of Assessment of Human Well-Being within a Socioecological Context: A Literature Review," *Social Indicators Research* 116, no. 3 (2014). On SDGs, see http://unstats.un.org/sdgs/. On the Gross National Happiness index, see http://www.gnhcentrebhutan.org/what-is-gnh/the-story-of-gnh/.

19. For background on the imitations of GDP, see Joseph E. Stiglitz, Amartya Sen, and Jean-Paul Fitousi, *Mismeasuring Our Lives: Why GDP Doesn't Add Up* (New York: The New Press, 2010).

20. Wackernagle and Rees, *Our Ecological Footprint*, 57 (italics added).

21. Cf. Norton, *Sustainability*, 304.

22. The common wisdom regarding debt is a staple of financial news and analysis: "For the best way to reduce sovereign debt, everyone agrees, is through economic growth. With growth, there are more jobs and more tax receipts, adding to government revenues, while the debt shrinks as a percentage of a rising gross domestic product." Steven Erlanger, "With Prospect of U.S. Slowdown, Europe Fears a Worsening Debt Crisis," *New York Times*, August 8, 2011, B3.

23. *Ponzi schemes* are a familiar form of pyramid scheme. Rather than actually investing the funds received from investors and allowing returns to fluctuate with investment performance, such schemes entice investors with false promises of consistently high returns and pay those returns directly from funds received from new investors while stealing the rest, until it becomes impossible to recruit new investors fast enough. In chapter 4, we will consider in some detail a social science model of pyramid-like dynamics of growth in social systems.

24. See Schor, *Plenitude: The New Economics of True Wealth*, 45ff, 73–75; Speth, *The Bridge at the Edge of the World: Capitalism, the Environment, and Crossing from Crisis to Sustainability*, 46, 55–57; and J. R. McNeill, *Something New under the Sun: An Environmental History of the Twentieth Century World* (New York: W. W. Norton, 2000).

25. James Speth, "The Limits of Growth," in *Moral Ground: Ethical Action for a Planet in Peril*, ed. Kathleen Moore and Michael P. Nelson (San Antonio, TX: Trinity University Press, 2010), 3, 6.

26. See Ahmet Atıl Aşıcı and Sevil Acar, "Does Income Growth Relocate Ecological Footprint?," *Ecological Indicators* 61 (2016); Thomas Dietz, Eugene A. Rosa, and Richard York, "Environmentally Efficient Well-Being: Is There a Kuznets Curve?," *Applied*

Geography 32, no. 1 (2012); Andrew K. Jorgenson and Thomas Dietz, "Economic Growth Does Not Reduce the Ecological Intensity of Human Well-Being," *Sustainability Science* 10, no. 1 (2015); and David Stern, "The Environmental Kuznets Curve after 25 Years," CCEP Working Paper 1514, Centre for Climate Economics and Policy, Crawford School of Public Policy, Australian National University, December 2015, https://ccep.crawford.anu.edu.au/sites/default/files/publication/ccep_crawford_anu_edu_au/2016-01/ccep1514_0.pdf.

27. This definition is compatible with the possibility of elements in a society qualifying as conducive to sustainability only with respect to the sustainability of a different society.

28. Such comparative judgments of the impact of decisions on ecosystem services are being made using tools for mapping and evaluating natural assets. See the Natural Capital Project homepage, http://www.naturalcapitalproject.org/, and Gretchen Daily et al., "Ecosystem Services in Decision Making: Time to Deliver," *Frontiers in Ecology and the Environment* 7, no. 1 (2009). For comparative analyses of the sustainability conduciveness of aspects of urban design, see Patrick Condon, *Seven Rules for Sustainable Communities: Design Strategies for the Post-Carbon World* (Washington, DC: Island Press, 2010).

29. The defining conditions for a free market require many independent suppliers and customers, whereas oligopolistic markets have only a few independent suppliers, like the markets in many kinds of products and services today.

30. Joseph Tainter, *The Collapse of Complex Societies* (Cambridge: Cambridge University Press, 1988), 4.

31. See Tainter, *Collapse of Complex Societies*; Charles Redman, *Human Impact on Ancient Environments* (Tucson: University of Arizona Press, 1999); Jared Diamond, *Collapse: How Societies Choose to Fail or Succeed* (New York: Viking, 2005); Patricia McAnany and Norman Yoffee, eds., *Questioning Collapse: Human Resilience, Ecological Vulnerability, and the Aftermath of Empire* (Cambridge: Cambridge University Press, 2010).

32. See Thomas Dietz, Eugene A. Rosa, and Richard York, "Driving the Human Ecological Footprint," *Frontiers in Ecology and the Environment* 5 (2007); and Walter Dodds, *Humanity's Footprint: Momentum, Impact, and Our Global Environment* (New York: Columbia University Press, 2008).

33. On simplification and the centrality of energy in sustaining sociopolitical complexity, see Tainter, *Collapse of Complex Societies*, 193, 197–99, 209–216.

34. WCED (World Commission on Environment and Development), *Our Common Future* (Geneva: United Nations, 1987), 12.

35. See James Speth and Peter Haas, *Global Environmental Governance* (Washington, DC: Island Press, 2006), 56–61.

36. Ibid., 59.

37. For detailed documentation of this claim, see Schor, *Plenitude*, chap. 3.

38. For a long-term view, see McNeill, *Something New under the Sun*. For an overview of the acceleration of consumption since 1980, see Schor, *Plenitude*, chap. 2.

39. Anthony D. Barnosky et al., "Introducing the Scientific Consensus on Maintaining Humanity's Life Support Systems in the 21st Century: Information for Policy Makers," *The Anthropocene Review* 1, no. 1 (2014).

40. Barnosky et al., "Introducing the Scientific Consensus," 79 (italics in original).

41. On the World Wildlife Fund and its *Living Planet Reports*, see http://wwf.panda .org/about_our_earth/all_publications/living_planet_report/. See also World Wildlife Fund, *Living Planet Report 2014*, 45.

42. See World Wildlife Fund, *Living Planet Report 2014*; and World Wildlife Fund, *Living Blue Planet Report 2015* (Gland, Switzerland: WWF International, 2015), 2, http:// www.worldwildlife.org/publications/living-blue-planet-report-2015. For region-by-region overviews, see UNEP, *Summary of the Sixth Global Environment Outlook, GEO-6, Regional Assessments: Key Findings and Policy Messages* (Nairobi: United Nations Environmental Programme, 2016), http://www.unep.org/publications/.

43. World Wildlife Fund, *Living Planet Report 2014*, 4, 8–9, 16, and 44–51.

44. Quoted in Steve Connor, "The State of the World? It Is on the Brink of Disaster," *Independent*, March 30, 2005, http://www.independent.co.uk/news/science/the-state -of-the-world-it-is-on-the-brink-of-disaster-530432.html.

45. Dodds, *Humanity's Footprint*, 12–16.

46. World Wildlife Fund, *Living Planet Report 2014*, 54. The UN's recently revised projection of 9.6 billion is 0.3 B higher than its previous estimate.

47. Dodds, *Humanity's Footprint*, 19–20.

48. Worldwatch Institute, *Vital Signs 2012* (Washington, DC: Island Press, 2012), 76–77. Molecule for molecule, methane is a GHG about twenty-nine times more potent than carbon dioxide, and nitrous oxide is about three hundred times more potent.

49. Worldwatch Institute, *Vital Signs 2012*, 44–45.

50. World Wildlife Fund, *Living Planet Report 2012*, 85–86.

51. UNEP. *Global Environment Outlook 5* (Valletta, Malta: Progress Press, Ltd., 2012), 46–47, http://www.unep.org/geo/pdfs/geo5/GEO5_report_full_en.pdf.

52. Ibid., 51–52, 112, 119–120, passim. Aquatic dead zones extend along the North American Atlantic and Gulf of Mexico coasts and include an 8,500-square-mile expanse of the Gulf of Mexico.

53. World Wildlife Fund, *Living Planet Report 2012*, 76.

54. UNEP, *Global Environment Outlook 5*, 73–74.

55. FAO, *World Review of Fisheries and Aquaculture* (Rome: FAO Fisheries Department, 2010), http://www.fao.org/docrep/013/i1820e/i1820e01.pdf; R. A. Myers and B. Worm, "Rapid Worldwide Depletion of Predatory Fish Communities," *Nature* 423 (2003); Boris Worm et al., "Impacts of Biodiversity Loss on Ocean Ecosystem Services," *Science* 3, no. 5800 (November 2006).

56. UNEP, *Global Environment Outlook 5*, 76.

57. Andrea Thompson, "CO2 Nears Peak: Are We Permanently above 400 PPM?," *Climate Central*, May 16, 2016, http://www.climatecentral.org/news/co2-are -we-permanently-above-400-ppm-20351. The National Oceanic and Atmospheric Administration (NOAA) datasets are available at CO2 Now: http://co2now.org/ Current-CO2/CO2-Now/Current-Data-for-Atmospheric-CO2.html. See also Potsdam Institute for Climate Impact Research and Climate Analysis, *Turn Down the Heat: Why a 4°C Warmer World Must Be Avoided* (Washington, DC: World Bank, 2013), xiv.

58. "Global Carbon Emissions Reach Record 10 Billion Tons, Threatening 2 Degree Target," *Science Daily*, December 6, 2011, http://www.sciencedaily.com/releases/ 2011/12/111204144648.htm.

59. See, e.g., Rockström et al., "A Safe Operating Space for Humanity."

60. H. Damon Matthews and Ken Caldeira, "Stabilizing Climate Requires Near-Zero Emissions," *Geophysical Research Letters* 35, no. LO4705 (2008), doi: 10.1029/ 2007GL032388; Juliette Eilperin, "Carbon Output Must Near Zero to Avert Danger, New Studies Say," *Washington Post*, March 10, 2008, A01.

61. See http://climate.nasa.gov/scientific-consensus/.

62. The IPCC is a scientific and member government body established by the World Meteorological Organization and the United Nations Environmental Programme, charged with providing periodic assessments of the state of climate science and knowledge of climate change. For a summary of AR5's findings, including a report addressed to policymakers, see http://www.ipcc.ch/report/ar5. Although the assessments represent the work of thousands of scientists, the role of member governments in approving assessment reports injects some politically motivated dilution of the strongest conclusions reached on the basis of the science. For detailed histories of industry-funded efforts to obscure the degree of scientific consensus concerning climate disruption, see Naomi Oreskes and Eric Conway, *Merchants of Doubt* (New York: Bloomsbury Press, 2010); and James Powell, *The Inquisition of Climate Science* (New York: Columbia University Press, 2011). Oreskes and Conway provide a detailed history of the evolution of industry resistance to public health and environmental regulation into a full-blown, orchestrated assault on science (especially epidemiology, climatology, and environmental science generally).

63. Fred Pearce, "UN Climate Report Is Cautious on Making Specific Predictions," *Environment 360*, March 24, 2014, http://e360.yale.edu/feature/un_climate_report _is_cautious_on_making_specific_predictions/2750/.

64. See the ICCP official slide show, http://www.ipcc.ch/report/ar5/syr/. For more recent studies on ice sheet collapse and ocean levels, see James Hansen et al., "Ice Melt, Sea Level Rise and Superstorms: Evidence from Paleoclimate Data, Climate Modeling, and Modern Observations that 2 °C Global Warming Could be Danger-ous," *Atmospheric Chemistry and Physics* 16 (March 22, 2016): 3761–3812, http:// www.atmos-chem-phys.net/16/3761/2016/acp-16-3761-2016.html; Jeff Tollefson, "Antarctic Model Raises Prospect of Unstoppable Ice Collapse," *Nature* 531, no. 7596 (March 31, 2016); and David Pollard and Robert M. DeConto, "Contribution of Antarctica to Past and Future Sea-Level Rise," *Nature* 531, no. 7596 (March 31, 2016) http://www.nature.com/nature/journal/v531/n7596/full/nature17145.html.

65. See http://www.ipcc.ch/pdf/ar5/UN_SG_statement_SYR_press_conference.pdf.

66. Potsdam Institute, *Turn Down the Heat*, xiii–iv, 13–18, 37–41. This report was commissioned by the World Bank as a basis for its future development funding, sig-naling a major reorientation of its economic development policies to focus on clean, efficient, and resilient initiatives. The World Bank has simultaneously launched a Wealth Accounting and Valuation of Ecosystem Services (WAVES; wavespartner-ship.org) program to help countries integrate valuations of natural capital into their development planning.

67. Potsdam Institute, *Turn Down the Heat*, xviii.

68. Dodds, *Humanity's Footprint*, 42 and 70–77.

69. *Global Environment Outlook 5*, 104–105; Jessica Kraft, "Running Dry," *Earth Island Journal* 28, no. 1 (Spring 2013).

70. Joseph Tainter and Tadeusz Patzek, *Drilling Down: The Gulf Oil Debacle and Our Energy Dilemma* (New York: Springer, 2012), 35. David Goodstein, a professor of applied physics and former vice provost at the California Institute of Technology, writes similarly that "our way of life, firmly rooted in the myth of an endless supply of cheap oil, is about to come to an end" (Goodstein, *Out of Gas: The End of the Age of Oil* [New York: W. W. Norton, 2004], 120.) One of Goodstein's supporting argu-ments is that coal, though abundant by volume, is a far less concentrated energy source than oil.

71. Tainter and Patzek, *Drilling Down*, 4, 22, 41–47, passim.

72. Ibid., 200.

73. Richard A. Kerr, "Natural Gas from Shale Bursts onto the Scene," *Science* 328, no. 5986 (June 25, 2010); Mark A. Latham, "BP Deepwater Horizon: A Cautionary Tale for CCS, Hydrofracking, Geoengineering and Other Emerging Technologies with

Environmental and Human Health Risks," *William and Mary Environmental Law and Policy Review* 36, no. 1 (2011); and Deyi Hou, Jian Luo, and Abir Al-Tabbaa, "Shale Gas Can Be a Double-Edged Sword for Climate Change," *Nature Climate Change* 2 (June 2012).

74. Kerr, "Natural Gas from Shale."

75. J. David Hughes, "A Reality Check on the Shale Revolution," *Nature* 494, no. 7437 (2013).

76. Daniel P. Schrag, "Is Shale Gas Good for Climate Change?" *Daedalus* 141, no. 2 (2012).

77. David MacKay, *Sustainable Energy—without the Hot Air* (Cambridge: UIT, 2009); and Tainter and Patzek, *Drilling Down*, 23, 29, 126, 195–196, 202–203, 207, 211. See also Jefferson W. Teste et al., *Sustainable Energy: Choosing among Options*, 2nd ed. (Cambridge, MA: MIT Press, 2012).

78. Tainter and Patzek, *Drilling Down*, 195.

79. Ibid.

80. Ibid., 196.

81. Worldwatch Institute, *Vital Signs, Vol 22: The Trends That Are Shaping Our Future* (Washington, DC: Worldwatch Institute, 2015); Eduardo Porter, "How Renewable Energy Is Blowing Climate Change Efforts Off Course," *New York Times*, July 19, 2016. http://www.nytimes.com/2016/07/20/business/energy-environment/how-renewable-energy-is-blowing-climate-change-efforts-off-course.html. Porter writes that renewable generating capacity in California and elsewhere is producing midday power gluts that the systems are not yet equipped to handle. He notes correctly that it is not enough to encourage a rapid growth of renewable generating capacity, but is incorrect in implying that such growth is not essential.

82. The net energy loss on corn ethanol, when all energy inputs are accounted for, has been calculated at 29 percent, and the loss on switch grass ethanol, if that became feasible, at 50 percent. Biomass for ethanol production is thus a source of liquid fuel but not a *source of energy* in the all-important sense of adding to humanity's usable supplies of energy. Even if there were a net energy gain, the vision of large-scale use of biofuels is fundamentally misguided. Photosynthesis is such an inefficient way of capturing the energy of sunlight (0.01 percent versus 2 percent for photovoltaic cells) that it would require an impractical diversion of biomass even to reach President George W. Bush's goal of thirty-five billion gallons of ethanol annually or to replace a mere 15 percent of US gasoline consumption by 2017. See David Pimentel and Tad Patzek, "Ethanol Production Using Corn, Switchgrass, and Wood; Biodiesel Production Using Soybean and Sunflower," *Natural Resources Research* 14, no. 1 (March 2005); UNEP, *Global Environment Outlook* 5, 68–71; and George Bush, "State of the Union Address," Washington, DC, 2007. By 2010, 29 percent of grain

production in the United States was already devoted to ethanol fuel production (Worldwatch Institute, *Vital Signs 2012*, 82), and projections for 2012, following a summer of severe drought, were for 26 percent of corn production yielding about 10.5 billion gallons of ethanol (International Energy Agency, "US Ethanol Production Plunges to Two-Year Low," IEA.org, August 13, 2012, https://www.iea.org/newsroomandevents/news/2012/august/us-ethanol-production-plunges-to-two-year-low.html.

83. David Goodstein estimates that known reserves of uranium for conventional power reactors would be enough to supply current energy demands for only five to twenty-five years. It is possible, in principle, to convert other, more abundant radioactive materials into usable fuel for reactors, but the risks are far beyond those of the nuclear technologies in commercial use. See Goodstein, *Out of Gas*, 106–107. This suggests that investment in nuclear energy is at best a short-term solution requiring exceedingly long-term investments in securing radioactive waste against leakage.

84. Tainter, *Collapse of Complex Societies*, 4.

85. Hansen et al., "Ice Melt, Sea Level Rise and Superstorms"; Tollefson, "Antarctic Model Raises Prospect of Unstoppable Ice Collapse"; Pollard and DeConto, "Contribution of Antarctica to Past and Future Sea-Level Rise."

86. Bjørn Lomborg, *The Skeptical Environmentalist* (Cambridge: Cambridge University Press, 2001), 70ff., 350–352.

87. Cass Sunstein, *Worst-Case Scenarios* (Cambridge, MA: Harvard University Press, 2007), 190.

88. Ibid.

89. Redman, *Human Impact on Ancient Environments*; McAnany and Yoffee, *Questioning Collapse*.

90. John Cooper, *Plato: Complete Works* (Indianapolis: Hackett, 1997), 1163–1164 [*Republic*, 552].

2 Obstacles to Sustainability

1. A United Nations Environmental Programme 2011 report estimated that economic losses owing to environmental damage totaled $6.6 trillion or 11 percent of global production in 2008, and that such losses could reach $28.6 trillion by 2050 in a business-as-usual scenario (UNEP FI, *Universal Ownership: Why Environmental Externalities Matter to Institutional Investors* [Geneva: PRI Association and UN Environmental Programme Finance Initiative, 2011], 88). A landmark 2006 report by the Treasury of the United Kingdom quantified the costs owing to climate change, projecting a 20 percent decline in global average consumption in a business-as-usual

scenario (Nicholas Stern, *Stern Review on the Economics of Climate Change* [London: HM Treasury, 2006], http://webarchive.nationalarchives.gov.uk/+/http:/www.hm -treasury.gov.uk/sternreview_index.htm). For estimates of costs of climate disruption already occurring, see NRDC, "Groundbreaking Study Quantifies Health Costs of U.S. Climate Change-Related Disasters & Disease," NRDC, November 8, 2011, http://www.nrdc.org/health/climate/extreme-weather-ticker-2012.asp; and World Wildlife Fund, "2012 Weather Extremes: Year-to-Date Review," December 6, 2012, http://www.wwfblogs.org/climate/sites/default/files/2012-Weather-Extremes-Fact -Sheet-6-dec-2012-final.pdf. The giant reinsurance company, Munich Re, maintains a NatCatSERVICE database that is widely cited as confirming that "the incidence of weather-related natural disasters since 1980 in Europe has more than doubled" (Wolfgang Kron, "Increasing Weather Losses in Europe: What They Cost the Insurance Industry?," *CESifo Forum* 12, no. 2 [2011], 74). For an analysis of which extreme weather events can be attributed to human-induced climate destabilization, see Stephanie C. Herring, Martin P. Hoerling, Thomas C. Peterson, and Peter A. Scott, eds., "Explaining Extreme Events of 2013 From a Climate Perspective," supplement, *Bulletin of the American Meteorological Society* 95, no. 9 (September 2014), http:// journals.ametsoc.org/doi/pdf/10.1175/1520-0477-95.9.S1.1. See also Gernot Wagner and Martin L. Weitzman, *Climate Shock: The Economic Consequences of a Hotter Planet* (Princeton, NJ: Princeton University Press, 2015).

2. Cf. Dodds, *Humanity's Footprint*, which offers a biological perspective on competitive accumulation and propensities to cooperate; Robert Gifford, "The Dragons of Inaction: Psychological Barriers That Limit Climate Change Mitigation," *American Psychologist* 66, no. 4 (2011) (examined ahead); Tim Kasser et al., "Materialistic Values: Their Causes and Consequences," in *Psychology and Consumer Culture: The Struggle for a Good Life in a Materialistic World*, ed. Tim Kasser and Allen Kanner (Washington, DC: American Psychological Association, 2004), which attributes materialism to social models and experiences that cause anxiety or personal insecurity; Elinor Ostrom, "A Multi-Scale Approach to Coping with Climate Change and Other Collective Action Problems," *Solutions* 1, no. 2 (February 2010), http://www .thesolutionsjournal.com/print/565, and related works discussed ahead; James Speth, *The Bridge at the Edge of the World* and *America the Possible: Manifesto for a New Economy* (New Haven, CT: Yale University Press, 2012), which focus on the dynamics of global capitalism and American political economy; Andrew Szasz, "Is Green Consumption Part of the Solution?," in *The Oxford Handbook of Climate Change and Society*, ed. John S. Dryzek, Richard B. Norgaard, and David Schlosberg (Oxford: Oxford University Press, 2011).

3. John Cavanaugh and Jerry Mander, *Alternatives to Global Capitalism: A Better World Is Possible* (San Francisco: Berrett-Koehler, 2002), 124. Proposals for reforms of corporate law that might be conducive to sustainability include reintroducing mandates to serve defined public purposes, restricting corporate political activity, and expanding the personal liability of corporate management and shareholders. Lynn

Stout, an influential scholar of US corporate law, has made the important argument that there is no existing legal basis for assertions that for-profit corporations have a singular duty to "maximize shareholder wealth." She offers a theory of the dependence of corporate activity on community partners that would justify sustainability-conducive reforms. See Stout, *The Shareholder Value Myth: How Putting Shareholders First Harms Investors, Corporations, and the Public* (San Francisco: Berrett-Koehler Publishers, 2012).

4. On the range of ways in which NGOs and SMOs facilitate collective political and social action to mitigate climate disruption and promote sustainability, see Ronnie Lipschutz and Corina Mckendry, "Social Movements and Global Civil Society," in *The Oxford Handbook of Climate Change and Society*, edited by John S. Dryzek, Richard B. Norgaard, and David Schlosberg (Oxford: Oxford University Press, 2011).

5. For a more elaborate overview of mechanisms of coordination, see John Dryzek, *Rational Ecology: Environment and Political Economy* (Oxford: Blackwell, 1987). See also Dryzek, *The Politics of the Earth: Environmental Discourses*, 3rd ed. (New York: Oxford University Press, 2013).

6. As we noted in chapter 1.

7. For related critiques of the IPAT formula and focus on individual consumer choices, see Thomas Princen, Michael Manites, and Ken Conca, eds., *Confronting Consumption* (Cambridge, MA: MIT Press, 2002), esp. chaps. 1, 3, 4, and 5.

8. Thomas Dietz et al., "Household Actions Can Provide a Behavioral Wedge to Rapidly Reduce US Carbon Emissions," *Proceedings of the National Academy of Sciences* 106, no. 44 (November 2009): 18452, 18453.

9. See Szasz, "Is Green Consumption Part of the Solution?"; and Frances Bowen, *After Greenwashing: Symbolic Corporate Environmentalism and Society* (Cambridge: Cambridge University Press, 2014).

10. Szasz, "Is Green Consumption Part of the Solution?," 601.

11. This knowledge is expressible as a *life cycle assessment* (LCA), which considers the environmental impacts associated with every stage and aspect of a product's origination, use, and disposal. For details, see the US EPA's LCA webpage, http://www.epa.gov/nrmrl/std/lca/lca.html#define.

12. As we discuss ahead, negative *externalities* (losses borne by third parties) are not reflected in price, and the UNEP has estimated the annual environmental externalities associated with the operations of the world's three thousand largest businesses at $2.1 trillion (UNEP FI, *Universal Ownership*, 88).

13. We address the nature and ethics of transparency in chapter 3.

14. Subsidized insurance is another sphere of economic transactions in which misalignment of price and environmental risk sometimes invites unsustainable choices.

Homeowner flood insurance is a prime example. See A. D. Eastman, "The Home-owner Flood Insurance Affordability Act: Why the Federal Government Should Not Be in the Insurance Business," *American Journal of Business and Management* 4, no. 2 (2015); A. B. McDonnell, "The Biggert-Waters Flood Insurance Reform Act of 2012: Temporarily Curtailed by the Homeowner Flood Insurance Act of 2014—A Respite to Forge an Enduring Correction to the National Flood Insurance Program Built on Virtuous Economic and Environmental Incentives," *Washington University Journal of Law and Policy* 49, no. 1 (2015); and Committee on the Affordability of National Insurance Program Premiums, *Affordability of National Flood Insurance Program Premiums. Report 1* (Washington, DC: National Academies Press, 2015), http://www.nap .edu/catalog/21709/affordability-of-national-flood-insurance-program-premiums -report-1.

15. Marilyn Brown Jess Chandler, Melissa V. Lapsa, and Benjamin K. Sovacool, *Carbon Lock-In: Barriers to Deploying Climate Change Mitigation Technologies* (Oak Ridge, TN: Oak Ridge National Laboratory, 2007), ix, xii. Support systems for incum-bent technologies would include government subsidies. See Jordan Weissmann, "America's Most Obvious Tax Reform: Kill the Oil and Gas Subsidies," *Atlantic*, March 19, 2013, http://www.theatlantic.com/business/archive/2013/03/americas -most-obvious-tax-reform-idea-kill-the-oil-and-gas-subsidies/274121/; and Oil Change International, "Fossil Fuel Subsidies: Overview," 2016, http://priceofoil.org/fossil-fuel -subsidies/.

16. We will not address the details of these approaches, except to observe that they would be economically equivalent; the same price on carbon can be set through an emissions allowance fee as through a tax. Where a cap-and-trade approach has been politically favored, evidently it has been to avoid using the word *tax*.

17. The *ethical* logic of calling this "fair" is not that prices can be put on all forms of third-party harm, including premature death, such that they could be fully compen-sated financially. It is rather that the price mechanism could function to limit use of products that impose risk of premature death to uses that are comparably vital to human well-being in the aggregate. A carbon tax or permit ideally would be priced to restrain the aggregate risk of premature death and catastrophic harm within pub-licly acceptable limits—balancing the harms resulting from restraining carbon emis-sions against commensurable harms resulting from higher carbon emissions—and this ideally would be determined through well-informed democratic processes. Note, by way of contrast, the choice of indoor smoking bans to prevent third parties from involuntarily suffering premature death from secondhand smoke; there is no inter-est at stake in smoking in public places that is remotely comparable to involuntary premature death, so the idea of "efficiently" pricing the premature deaths into smoking in public places is ethically specious. A ban is the ethically defensible choice. For an examination of some of the issues at stake, see Robert Goodin, "Sell-ing Environmental Indulgences," in *Climate Ethics: Essential Readings*, ed. Stephen M. Gardiner et al. (Oxford: Oxford University Press, 2010).

18. The idea of a society being well-ordered with respect to principles of justice and evidence is explained in chapter 4.

19. We owe this delightful formulation to Judith Lichtenberg, "Consuming because Others Consume," *Social Theory and Practice* 22, no. 3 (Fall 1996): 277. See also Jared Diamond, "Invention Is the Mother of Necessity," *New York Times Magazine*, 1999, http://partners.nytimes.com/library/magazine/millennium/m1/diamond .html. Diamond writes that "invention is the mother of necessity, by creating needs that we never felt before," but he does not distinguish mere felt needs from necessities (4).

20. Adam Smith, *Wealth of Nations*, ed. R. H. Campbell and Andrew S. Skinner (Oxford: Clarendon Press, 1976), vol. 2, 469–471. For discussion, see Lichtenberg, "Consuming because Others Consume," 282ff.; Amartya Sen, *Development as Freedom* (Oxford: Oxford University Press, 1999), esp. 73–74.

21. Lichtenberg, "Consuming because Others Consume," 284.

22. On the impact of social position on well-being, see Schor, *Plenitude*, esp. 176ff.; and Brian Barry, *Why Social Justice Matters* (Cambridge: Polity Press, 2005), esp. parts 2 and 5. For overviews of research on the unimportance of absolute affluence to happiness above a threshold of poverty, see Tim Kasser, *The High Price of Materialism* (Cambridge, MA: MIT Press), 43ff.; and Speth, *The Bridge at the Edge of the World*, 126–146.

23. Diamond, *Collapse*, 211–276, and esp. 222–230. The Norse settlement in Greenland was established in 984 and survived until sometime in the 1400s.

24. Jane Perlez and Kirk Johnson, "Behind Gold's Glitter: Torn Lands and Pointed Questions," *New York Times*, October 24, 2005, http://www.nytimes.com/2005/10/ 24/international/24GOLD.html?th+&emc+th&pagewa; and Jane Perlez and Lowell Bergman, "Tangled Strands in Fight over Peru Gold Mines," *New York Times*, October 25, 2005, http://www.nytimes.com/2005/10/25/international/americas/25GOLD. html?th+&emc=th. The damage caused by gold panning is in some respects worse, since it commonly involves direct handling and spilling of mercury into streams.

25. Brook Larmer, "The Real Price of Gold," *National Geographic* 215, no. 1 (January 2009).

26. Diamond, *Collapse*, 79–119, details evidence of deforestation that was all but complete by 1400 to 1600 and an associated population collapse of 70 percent by the 1700s from its peak around 1400–1600. However, see also Patricia McAnany and Norman Yoffee, eds., *Questioning Collapse: Human Resilience, Ecological Vulnerability, and the Aftermath of Empire* (Cambridge: Cambridge University Press, 2010), for correction of details, especially regarding the role of stowaway rats—an invasive species unwittingly brought to the island—in severely limiting forest regeneration.

27. See Kasser, *The High Price of Materialism*; Kasser et al., "Materialistic Values"; and Tim Kasser et al., "Some Costs of American Corporate Capitalism: A Psychological Exploration of Value and Goal Conflicts," *Psychological Inquiry* 18, no. 1 (2007).

28. See Juliet Schor, *Born to Buy: The Commercialized Child and the New Consumer Culture* (New York: Scribner, 2004); and Kasser et al., "Some Costs of American Corporate Capitalism."

29. Laurie Mazur and Shira Saperstein, "Afterward: Work for Justice?," in *A Pivotal Moment: Population, Justice & The Environmental Challenge*, ed. Laurie Mazur (Washington, DC: Island Press, 2010), 394. For an earlier work that does discuss connections between population and environment, see B. O'Neill, F. L. MacKellar, and W. Lutz, *Population and Climate Change* (Cambridge: Cambridge University Press, 2001).

30. UNFPA, *Programme of Action: Adopted at the International Conference on Population and Development, Cairo, 5–13 September 1994* (Geneva: United Nations Population Fund, 2004), 46.

31. Jacqueline Nolley Echegaray and Shira Saperstein, "Reproductive Rights Are Human Rights," in *A Pivotal Moment*, ed. Mazur, 348; Mazur, ed., *A Pivotal Moment*, 395; Martha Nussbaum, "Women's Education: A Global Challenge," *Signs* 29, no. 2 (2003); and Sen, *Development as Freedom*, 195–199. Sen writes: "The negative linkage between female literacy and fertility appears to be, on the whole, empirically well founded. Such connections have been widely observed ... The unwillingness of educated women to be shackled to continuous child rearing clearly plays a role ... Education also makes the horizon of vision wider, and ... helps to disseminate the knowledge of family planning. And of course educated women tend to have greater freedom to exercise their agency in family decisions, including in matters of fertility and childbirth" (199).

32. Jeffrey D. Sachs, *The Age of Sustainable Development* (New York: Columbia University Press, 2015), 253–254.

33. See Philip Kitcher, "Public Knowledge and Its Discontents," *Theory and Research in Education* 9, no. 2 (2011); and Kitcher, *Science in a Democratic Society* (Amherst, NY: Prometheus Books, 2011).

34. See, in addition to Kitcher's works cited in note 33, Allen Buchanan, "Social Moral Epistemology," *Social Philosophy and Policy* 19 (2002); Buchanan, "Political Liberalism and Social Epistemology," *Philosophy & Public Affairs* 32, no. 2 (2004); and Howard Gardner, Mihaly Csikszentmihalyi, and William Damon, *Good Work* (New York: Basic Books, 2001), which addresses both research and journalism at length.

35. Noel Castree, "Reply to 'Strategies for Changing the Intellectual Climate' and 'Power in Climate Change Research,'" *Nature Climate Change* 5 (May 2015). See Noel

Castree et al., "Changing the Intellectual Climate," *Nature Climate Change* 4 (September 2014); Myanna Lahsen et al. "Strategies for Changing the Intellectual Climate," *Nature Climate Change* 5 (May 2015); and Lauren Rickards, "Power in Climate Change Research," *Nature Climate Change* 5 (May 2015). For an alternative conception of how science may adequately inform and be informed by ethical and political considerations, see Kitcher, "Public Knowledge and Its Discontents" and *Science in a Democratic Society*. See also the opening of chapter 4.

36. For a manifesto on behalf of integrative science, see Murray Gell-Mann, "Transformations of the Twenty-First Century: Transitions to Greater Sustainability," in *Global Sustainability: A Nobel Cause*, ed. Hans Joachim Schellnhuber et al. (Cambridge: Cambridge University Press, 2010).

37. See James Powell, *The Inquisition of Climate Science* (New York: Columbia University Press, 2011), esp. chapter 11, "Balance as Bias"; Naomi Oreskes and Erik Conway, *Merchants of Doubt*, esp. the epilogue, "A New View of Science"; Riley E. Dunlap and Aaron M. McCright, "Organized Climate Change Denial," in *The Oxford Handbook of Climate Change and Society*, ed. John S. Dryzek, Richard B. Norgaard, and David Schlosberg; and Judith A. Layzer, *Open for Business: Conservatives' Opposition to Environmental Regulation* (Cambridge, MA: MIT Press, 2014).

38. See Anthony Leiserowitz et al., *Climate Change in the American Mind: October 2015* (New Haven, CT: Yale Project on Climate Change Communication and George Mason University Center on Climate Change Communication, 2015), http://climatecommunication.yale.edu/wp-content/uploads/2015/11/Climate -Change-American-Mind-October-20151.pdf. Although two-thirds of Americans understand that global warming is happening and half understand it is mostly human caused, only about 10 percent realize that nearly all climatologists accept that human-caused climate change is happening. A related spring 2016 study by the same authors (*Global Warming and the U.S. Presidential Election, Spring 2016* [New Haven, CT: Yale Project on Climate Change Communication and George Mason University Center on Climate change Communication, 2016], http:// climatecommunication.yale.edu/wp-content/uploads/2016/05/2016_3_CCAM _Global-Warming-U.S.-Presidential-Election.pdf) found that a majority of the supporters of each major party presidential candidate other than Ted Cruz believe global warming is occurring, but the study also confirmed previous findings that Americans' belief in and level of concern about climate change varies widely by major political party affiliation. Supporters' belief in global warming ranged from 38 percent (Cruz) and 56 percent (Trump) to 92 percent (Clinton) and 93 percent (Sanders). At the time of Leiserowitz et al.'s 2012 study of the political affiliation split, 81 percent of Democrats but only 47 percent of Republicans agreed that global warming is occurring (Anthony Leiserowitz et al., *Global Warming's Six Americas in March 2012 and November 2011* [New Haven, CT: Yale Project on Climate Change Communication and George Mason University Center for Climate Change Communication, 2012], http://environment.yale.edu/climate/files/Six-Americas

-March-2012.pdf). It is unlikely that gaps of the magnitude observed could exist without substantial variation in what media sources are trusted.

39. Bowen, *After Greenwashing*.

40. Elliott Negin, "Documenting Fossil Fuel Companies' Climate Deception," *Catalyst* 14 (Summer 2015). For the eighty-five internal industry documents and Union of Concerned Scientists report on which this article is based, see Union of Concerned Scientists, *The Climate Deception Dossiers* (Cambridge, MA: UCS, 2015), http://www.ucsusa.org/decadesofdeception.

41. For extensive documentation of these and related claims, see Kristin Schrader-Frechette, *Taking Action, Saving Lives* (New York: Oxford University Press, 2007); Oreskes and Conway, *Merchants of Doubt*; Powell, *Inquisition of Climate Science*; Dunlap and McCright, "Organized Climate Change Denial"; and Union of Concerned Scientists, *Climate Deception Dossiers*.

42. Powell, *Inquisition of Climate Science*, 94–95, 110.

43. Powell, ibid., 111. See also Union of Concerned Scientists, *Smoke, Mirrors, and Hot Air: How ExxonMobil Uses Big Tobacco's Tactics to Manufacture Uncertainty on Climate Science* (Cambridge, MA: UCS, 2007).

44. Union of Concerned Scientists, *Climate Deception Dossiers*, 11.

45. Kitcher, "Public Knowledge and Its Discontents," 119–120.

46. Powell, *Inquisition of Climate Science*, 11, quoting the keynote address, "Climate Alarm," by Richard Lindzen, an MIT professor of meteorology, speaking at a March 2009 Heartland Institute conference. The consensus seems to be that neither Lindzen's research nor anyone else's has identified any significant negative feedback mechanisms that would slow climate change, while several positive feedbacks are outpacing IPCC projections: the retreat of sea ice (causing a decline of surface reflectivity), accelerating decay of organic matter in soil (releasing GHGs), decline of forests killed by drought and pests, and release of methane from warming sea beds and melting permafrost. Droughts in the Amazon in 2005 and 2010 caused releases of carbon measured in gigatons (GtC, or billions of tons of carbon), estimated at 1.2 to 3.4 GtC in 2010, and reduced subsequent carbon sequestration. See WWF, *Living Planet Report 2012*, 95; and Brian Kahn, "Drought Weakens the Amazon's Ability to Capture Carbon," *Climate Central*, March 9, 2015, http://www.climatecentral.org/news/drought-amazon-carbon-capture-18733.

47. Gifford, "Dragons of Inaction." For investigations of the perception of risk in particular, see Robert E. O'Connor, Robert E., Richard J. Bird, and Ann Fisher, "Risk Perceptions, General Environmental Beliefs, and Willingness to Address Climate Change," *Risk Analysis* 19 (1999); Peter Slovic, *The Perception of Risk* (London: Earthscan, 2000); and Anthony Leiserowitz, "American Risk Perceptions: Is Climate Change Dangerous?," *Risk Analysis* 25 (2005).

48. For a comprehensive examination of the forms and dynamics of denial, see Stanley Cohen, *States of Denial* (Cambridge: Polity Press, 2001).

49. See Kari Marie Norgaard, *Living in Denial: Climate Change, Emotions, and Everyday Life* (Cambridge, MA: MIT Press, 2011), esp. 207ff.

50. Ibid., 5.

51. Cf. Ottar Helevik, "Beliefs, Attitudes, and Behavior towards the Environment," in *Realizing Rio in Norway: Evaluative Studies of Sustainable Development*, ed. William Laverty, Morton Nordskog, and Hilde Annette Aakre (Oslo: Program for Research and Documentation for a Sustainable Society, University of Oslo, 2002), 13.

52. Laurie Michaelis, "Consumption Behavior and Narratives about the Good Life," in *Creating a Climate for Change: Communicating Climate Change and Facilitating Social Change*, ed. Susanne Moser and Lisa Dilling (Cambridge: Cambridge University Press, 2007), 254.

53. Philosophers distinguish what is *irrational*, or contrary to what a person has good reason to believe (given the evidence available to her) or do (given her beliefs, commitments, and preferences), from what is *unreasonable*, or contrary to what one could justify to others as free and equal citizens of one's society or the world.

54. These difficulties, and proposals for how to overcome them in communications concerning climate change, are outlined in John D. Sterman, "Communicating Climate Change Risks in a Skeptical World," *Climatic Change* 108, no. 4 (2011), doi:10.1007/s10584-011-0189-3, http://link.springer.com/article/10.1007%2Fs10584-011-0189-3. Readers interested in a fuller accounting of general defects of individual rationality or judgment might begin with Thomas Gilovic, Dale Griffin, and Daniel Kahneman, *Heuristics and Biases: The Psychology of Intuitive Judgment* (Cambridge: Cambridge University Press, 2002); Daniel Kahneman and Amos Tversky, *Choices, Values and Frames* (Cambridge: Cambridge University Press, 2000); and D. Kahneman, P. Slovic and A. Tversky, *Judgment under Uncertainty* (Cambridge: Cambridge University Press, 1982).

55. See Kasser et al., "Materialistic Values"; and Dodds, *Humanity's Footprint*.

56. See Kris Kirby and R. J. Herrnstein, "Preference Reversals Due to Myopic Discounting of Delayed Rewards," *Psychological Science* 6 (1995); Andrew Millar and Douglas Navarick, "Self-Control and Choice in Humans," *Learning and Motivation* 15 (1984); Ted O'Donoghue and Matthew Rabin, "Doing It Now or Later," *American Economic Review* 89 (1999); George Ainslie, *Breakdown of Will* (Cambridge: Cambridge University Press, 2001); Chrisoula Andreou, "Understanding Procrastination," *Journal for the Theory of Social Behaviour* 37 (2007); and Andreou, "Environmental Preservation and Second-Order Procrastination," *Philosophy & Public Affairs* 35, no. 3 (2007).

57. See Kasser, *High Price of Materialism*.

58. See Jon Elster, *Ulysses Unbound: Studies in Rationality, Precommitment, and Constraints* (Cambridge: Cambridge University Press, 2000).

59. Garrett Hardin, "The Tragedy of the Commons," *Science* 162 (1968): 1243–1248.

60. Ostrom, "A Multi-Scale Approach," 29. See also Thomas Dietz, "Elinor Ostrom: 1933–2012," *Solutions* 3, no. 5 (August 2012), http://www.thesolutionsjournal.com/node/1166.

61. Much of the research is summarized in Committee on the Human Dimensions of Global Change et al., eds., *The Drama of the Commons* (Washington, DC: National Academies Press, 2002), http://www.nap.edu/openbook.php?record_id=10287 &page=1. See esp. chap. 2: Arun Agrawal, "Common Resources and Institutional Sustainability." See also Paul Stern, Thomas Dietz, and Elinor Ostrom, "Research on the Commons: Lessons for Environmental Resource Managers," *Environmental Practice* 4, no. 2 (June 2002).

62. Ostrom, "A Multi-Scale Approach," 29. Josiah Ober offers a similar account in *Democracy and Knowledge: Innovation and Learning in Classical Athens* (Princeton, NJ: Princeton University Press, 2008).

63. Elinor Ostrom, "A General Framework for Analyzing Sustainability of Social-Ecological Systems," *Science* 325 (July 24, 2009): 420–421. The discussion of leadership in these pages notes that "entrepreneurial skills" and "the presence of college graduates" have a "strong positive effect."

64. The ethics of legitimacy will be addressed in chapter 3. Ostrom writes: "Trust that government officials are objective, effective, and fair is more important in enabling a governmental policy to work than reliance on force" ("A Multi-Scale Approach," 29).

65. Mark Cooper, "The Economic and Institutional Foundations of the Paris Agreement on Climate Change: The Political Economy of Roadmaps to a Sustainable Electricity Future," January 26, 2016, http://papers.ssrn.com/sol3/Papers.cfm?abstract_id=2722880; Elinor Ostrom, "Polycentric Systems for Coping with Collective Action and Global Environmental Change," *Global Environmental Change* 20, no. 4 (2010); Ostrom, "Nested Externalities and Polycentric Institutions: Must We Wait for Global Solutions to Climate Change Before Taking Actions at Other Scales?," *Economic Theory* 49, no. 2 (2012); and Timothy Randhir, "Globalization Impacts on Local Commons: Multiscale Strategies for Socioeconomic and Ecological Resilience," *International Journal of the Commons* 10, no. 1 (2016), https://www.thecommonsjournal.org/article/10.18352/ijc.517/.

66. See Kent E. Portney, *Taking Sustainable Cities Seriously: Economic Development, the Environment, and Quality of Life in American Cities*, 2nd ed. (Cambridge, MA: MIT Press, 2013), chap. 5; and Layzer, *Open for Business*. Portney reviews the existing research on local environmental activism and concludes that it is most efficacious in

engendering cooperation (to overcome tragedy of the commons problems, overcome inequitable siting of environmental hazards, etc.) "when participation takes place in the context of local nonprofit organizations," is engaged with city government, and "performs administrative-type functions related to sustainability" (185). Progress toward sustainable communities will require "a great deal of attention to fostering community-building processes" (184). Layzer, too, notes limits to the "transformative power" of local environmental activism while also noting its attractions. Like many others, she emphasizes the importance of "policies that structure the incentives people face" (364). Randall Curren and Chuck Dorn address aspects of the community-building processes that concern Portney; see Curren and Dorn, *Patriotic Education in a Global Age* (Chicago: University of Chicago Press, 2017), chaps. 4 and 5. Their focus is on shared values, civic friendship, and the satisfaction of relational, competence, and self-determination needs that activism through nonprofit organizations can provide—both locally and globally.

67. See Lipschutz and Mckendry, "Social Movements and Global Civil Society"; M. E. Keck and K. Sikkink, *Activists beyond Borders: Advocacy Networks in International Politics* (Ithaca, NY: Cornell University Press, 1998); Kate Nash, "Towards Transnational Democratization?," in *Transnationalizing the Public Sphere*, ed. Kate Nash (Cambridge: Polity Press, 2014); J. Scholte, ed., *Building Global Democracy? Civil Society and Accountable Global Governance* (Cambridge: Cambridge University Press, 2011); and S. Tarrow, *The New Transnational Activism* (Cambridge: Cambridge University Press, 2005).

68. Karin Bäckstrand, "The Democratic Legitimacy of Global Governance after Copenhagen," in *The Oxford Handbook of Climate Change and Society*, ed. John S. Dryzek, Richard B. Norgaard, and David Schlosberg (Oxford: Oxford University Press, 2011).

69. Ostrom, "Nested Externalities and Polycentric Institutions."

70. See Daniel Farber, "Issues of Scale in Climate Governance," in *The Oxford Handbook of Climate Change and Society*, ed. John S. Dryzek, Richard B. Norgaard, and David Schlosberg (Oxford: Oxford University Press, 2011).

71. James Speth and Peter Haas, *Global Environmental Governance* (Washington, DC: Island Press, 2006). For another view of the mediating role that regional self-governance might play in building global environmental governance, see Ken Conca, "The Rise of the Region in Global Environmental Governance," *Global Environmental Politics* 12, no. 3 (August 2012).

72. William Nordhaus, "A New Solution: The Climate Club," *New York Review of Books* 52, no. 10 (June 4, 2015): 39. For a broad, rights-based approach to green governance that incorporates insights from Ostrom's work, see Burns H. Weston and David Bollier, *Green Governance: Ecological Survival, Human Rights, and the Law of the Commons* (Cambridge: Cambridge University Press, 2014).

3   Sustainability Ethics and Justice

1. The use of measures of subjective well-being in pursuit of sustainability is defended in Thomas Dietz, "Informing Sustainability Science through Advances in Environmental Decision Making and Other Areas of Science" (paper presented at the NRC Sustainability Science Roundtable, Irvine, CA, January 14–15, 2016), http://sites.nationalacademies.org/cs/groups/pgasite/documents/webpage/pga_170344.pdf.

2. See Kathleen D. Moore and Michael P. Nelson, eds., *Moral Ground: Ethical Action for a Planet in Peril* (San Antonio, TX: Trinity University Press, 2010); and Ryne Raffaelle, Wade Robison, and Evan Selinger, eds., *Sustainability Ethics: 5 Questions* (Copenhagen: Vince, Inc. Automatic Press, 2010). These edited volumes consist of many short essays by diverse writers, including such well-known figures as the Dalai Lama, Desmond Tutu, E. O. Wilson, and Ursula K. Le Guin. For a summary and quotes from Pope Francis's groundbreaking climate encyclical, see Andrea Thompson, "Pope's Climate Encyclical: 4 Main Points," *Climate Central*, June 18, 2015, http://www.climatecentral.org/news/4-main-points-pope-climate-encyclical-19129.

3. See, e.g., David Crocker and Toby Linden, eds., *Ethics of Consumption: The Good Life, Justice, and Global Stewardship* (Lanham, MD: Rowman & Littlefield, 1998); Brian Barry, "Sustainability and Intergenerational Justice," in *Environmental Ethics*, ed. Andrew Light and Holmes Rolston III (Malden, MA: Blackwell, 2003); Lisa Newton, *Ethics and Sustainability* (Upper Saddle River, NJ: Prentice-Hall, 2003); Bryan Norton, *Sustainability: A Philosophy of Adaptive Ecosystem Management* (Chicago: University of Chicago Press, 2005); Cass Sunstein, *Worst-Case Scenarios* (Cambridge, MA: Harvard University Press, 2007); Naomi Zack, *Ethics for Disaster* (Lanham, MD: Rowman & Littlefield, 2009); Peter G. Brown and Jeremy J. Schmidt, eds., *Water Ethics: Foundational Readings for Students and Professionals* (Washington, DC: Island Press, 2010); Stephen M. Gardiner et al., eds., *Climate Ethics: Essential Readings* (Oxford: Oxford University Press, 2010); Christian Becker, *Sustainability Ethics and Sustainability Research* (Dordrecht, Netherlands: Springer, 2011); John Broome, *Climate Matters: Ethics in a Warming World* (New York: Norton, 2012); Willis Jenkins, *The Future of Ethics: Sustainability, Social Justice, and Religious Creativity* (Washington, DC: Georgetown University Press, 2013); and Dale Jamieson, *Reason in a Dark Time* (Oxford: Oxford University Press, 2014).

4. Immanuel Kant, *Grounding for the Metaphysics of Morals*, trans. James Ellington (Indianapolis: Hackett, 1981); Barbara Herman, "Mutual Aid and Respect for Persons," in *Kant's Groundwork of the Metaphysics of Morals*, edited by Paul Guyer (Lanham, MD: Rowman & Littlefield, 1998); Onora O'Neill, "Consistency in Action," in *Kant's Groundwork of the Metaphysics of Morals*; and Onora O'Neill, "Constructivism in Rawls and Kant," in *The Cambridge Companion to Rawls*, ed. Samuel Freeman (Cambridge: Cambridge University Press, 2003).

5. For background on this duty of care in the American legal context, see G. Edward White, *Tort Law in America: An Intellectual History* (New York: Oxford University Press, 1980).

6. Kant, *The Metaphysics of Morals*, sec. 6.

7. This thesis is developed and defended at length in Randall Curren, *Aristotle on the Necessity of Public Education* (Lanham, MD: Rowman & Littlefield, 2000).

8. See Tom Tyler, *Why People Obey the Law* (Princeton, NJ: Princeton University Press, 2006). Its central findings are that people obey the law if they think it is legitimate and that enforcement of law plays a comparatively minor role.

9. We will note in passing that these initial principles rely on the idea of natural capital, defined as "natural assets" that yield flows of "valuable goods or services into the future" (Robert Costanza and Herman Daly, "Natural Capital and Sustainable Development," *Conservation Biology* 6, no. 1 (1992): 38), and this conceptualizes nature as having only instrumental value. We noted in chapter 1 that some contributions of nature to the quality of human life may depend on regarding it as having noninstrumental value, but from the point of view of preserving aggregate human opportunities to live well, this does not require the introduction of a further principle. Human opportunity to experience nature in meaningful ways would not trump other kinds of good opportunities as a matter of sustainability, unless the best understanding of human well-being requires it.

10. The role of judgment in applying this principle should not be supposed to be unique or objectionable. Principles play a number of important roles in guiding, judging, and justifying action, but it would be a mistake to think that there is no judgment involved in applying them.

11. WHO, *Climate Change and Health, Fact Sheet No. 266, Revised August* 2014 (Geneva: WHO Media Centre, 2014), http://www.who.int/mediacentre/factsheets/fs266/en/. In 2007, WHO's estimate of direct deaths owing to climate change was 150,000 per year. *Direct* means owing to heat waves, violent storms, and elevated incidence of disease and excludes indirect deaths owing to reduced availability of food or water. The 2014 projection for 2030 to 2050 is 250,000 excess direct deaths. See also Tee L. Guidotti, *Health and Sustainability: An Introduction* (Oxford: Oxford University Press, 2015).

12. See Amy Gutmann and Dennis Thompson, *Democracy and Disagreement* (Cambridge, MA: Harvard University Press, 1996), 95ff.

13. Civic and economic efficacy or efficiency in this sense does not require and would not license people knowing things they have no legitimate interest in knowing. Invasions of the privacy of natural persons to satisfy voyeuristic curiosity or illicit desire would not be licensed, nor would it require disclosure of information

that would predictably facilitate unjust acts. For instance, when there are asymmetries of bargaining position, such as in labor negotiations in oligopolistic markets, full transactional transparency would not require the disadvantaged party to reveal aspects of her person or circumstances that would predictably facilitate an unfair outcome. An example of such an outcome would be exclusion from employment owing to a disabling condition irrelevant to qualifications or probable job performance.

14. See Gutmann and Thompson, *Democracy and Disagreement*, 98.

15. Ibid., 104.

16. See Powell, *Inquisition of Climate Science*; Oreskes and Conway, *Merchants of Doubt*; Dunlap and McCright, "Organized Climate Change Denial"; Negin, "Documenting Fossil Fuel Companies' Climate Deception"; and Layzer, *Open for Business*.

17. See chapter 2, note 3, and related text.

18. Readers familiar with legal doctrine will recognize this principle as a moral version of the legal doctrine of detrimental reliance or reliance-based estoppel (equitable estoppel), more general than the legal doctrine with respect to the pathway to reliance, more specific with respect to the character of reliance, and lacking the requirement of an actionable or compensable harm. Morally, it is sensible to treat imposition of unreasonable risk per se as a wrong, our concern being to capture what is ethically distinctive about unsustainability.

19. In its application to such cases, the *principle of detrimental reliance* adds pointed specificity to the first principle's prohibition against diminishing natural capital, subject to qualifications associated with the third principle or requirement to seek fair terms of cooperation. Those qualifications are met in the poaching case unless there is reason to think the protection of territorial waters is so politically unjust as to have no moral significance.

20. James Garvey, *The Ethics of Climate Change* (London: Continuum, 2008), 85.

21. The account that follows is drawn from Timothy Egan, *The Worst Hard Time* (New York: Mariner Books, 2006).

22. Our Aristotelian perspective on the nature and formation of moral virtue is developed in Randall Curren, "Motivational Aspects of Moral Learning and Progress," *Journal of Moral Education* 43, no. 4 (December 2014); "Judgment and the Aims of Education," *Social Philosophy & Policy* 31, no. 1 (Fall 2014); "Virtue Ethics and Moral Education," in *Routledge Companion to Virtue Ethics*, ed. Michael Slote and Lorraine Besser-Jones (London: Routledge, 2015); and "A Virtue Theory of Moral Motivation" (paper presented at the Varieties of Virtue Ethics in Philosophy, Social Science and Theology Conference, Oriel College, Oxford, January 8–10, 2015), http://www.jubileecentre.ac.uk/userfiles/jubileecentre/pdf/conference-papers/Varieties_of_Virtue_Ethics/Curren_Randall.pdf.

23. See FitzPatrick, "Valuing Nature Non-Instrumentally." The significance of sapience or the capacity to think is that it is essential to consciously valuing anything, including oneself, the conditions of one's life, and so on. Being self-valuing may be the best test or best meaning we can give to the idea of intrinsic value.

24. The presuppositions, scope, and substance of citizenship are all matters of debate and have been addressed in connection with sustainability and the environment in such works as Andrew Dobson, *Citizenship and the Environment* (Oxford: Oxford University Press, 2003), and Andrew Dobson and Derek Bell, eds., *Environmental Citizenship* (Cambridge, MA: MIT Press, 2006). We follow Curren and Dorn, *Patriotic Education*, chap. 6, "Global Civic Education," in conceiving of global citizenship as participation in global constitutional activity, including global environmental governance, defining *constitutional activity* as activity whereby people constitute themselves as a functional and affective global public and shape and preserve the norms and constitutional principles that regulate the global order.

25. Melissa Lane, *Eco-Republic: What the Ancients Can Teach Us about Ethics, Virtue, and Sustainable Living* (Princeton, NJ: Princeton University Press, 2012), 170, 164. *Moderation* and *self-discipline* are alternative translations of the Greek word *sôphrosunê*.

26. Ibid., 164–165. Lane identifies two principal tasks or ethical responsibilities belonging to "the ethics of responsible initiative": "to press towards an understanding of the whole, and to press towards action in relation to it" (170, 171). Putting these together, action conducive to sustainability must be undertaken "in light of the largest possible framing and understanding of the good" (173). Both of these ethical responsibilities are implicit in the virtues we have enumerated.

27. Ibid., 183.

28. John Locke, *Second Treatise of Government* (Indianapolis: Hackett, 1980), chap. 5 ("Of Property"), esp. secs. 28 and 32–47. Because this Lockean proviso *has* been violated, one might ask whether it is not immoral, as Locke assumed, to deny some the means to provide for themselves and meet their needs so that others may accumulate wealth without end. Locke suggests (secs. 47–50) that the invention of money rendered the accumulation of wealth innocuous and may expand wealth and opportunity so far as to make the proviso irrelevant, but if this was his view, then it is still predicated on opportunity that is essentially unlimited or expanding without limit. We are grateful to Richard Dees for calling this to our attention.

29. Moses Finley, *Land and Credit in Ancient Athens, 500–200 B.C.* (New Brunswick, NJ: Rutgers University Press, 1953); and Finley, *Politics in the Ancient World* (Cambridge: Cambridge University Press, 1983).

30. Ryan Balot documents the widespread attention to these themes in Athens and their importance for understanding Plato's work in *Greed and Injustice in Classical Athens* (Princeton, NJ: Princeton University Press, 2001).

31. The translations from Plato's works relied on here are from John Cooper, ed., *Plato: Complete Works* (Indianapolis: Hackett, 1997). See *Republic*, 369–372, 372e, 373. Page references are to the "Stephanus numbers" of Henri Estienne's Greek text of Plato. These appear in the margins of most English translations.

32. *Republic*, 372d, 372d, 369dff., 372c, 372b–c, 372c.

33. Ibid., 372b–d.

34. See Clive Ponting, *A Green History of the World* (London: Penguin, 1991), 76–77; and Michael Williams, *Deforesting the Earth: From Prehistory to Global Crisis (An Abridgment)* (Chicago: University of Chicago Press, 2006), 62ff.

35. *Critias*, 111b.

36. Ibid., 117–121.

37. Ibid., 121a.

38. Aristotle, *Politics* 1256b27–1258b8, 1265a38–b16, 1326a5–b2, 1327a15, 1335b21–27, 1301a36–b4. These page references refer to the page, column, and line numbers of Immanuel Bekker's 1831 edition of the Greek text of Aristotle's works. These are the numbers that appear in the margins of most modern editions and translations.

39. See Richard M. Ryan, Randall Curren, and Edward L. Deci, "What Humans Need: Flourishing in Aristotelian Philosophy and Self-Determination Theory," in *The Best within Us: Positive Psychology Perspectives on Eudaimonia*, ed. Alan S. Waterman (Washington, DC: American Psychological Association, 2013); Richard M. Ryan, Veronika Huta, and Edward L. Deci, "Living Well: A Self-Determination Theory Perspective on Eudaimonia," *Journal of Happiness Studies* 9 (2008); and Kasser, *High Price of Materialism*.

40. Rawls contrasts the just *property-owning democracy* he defends with four kinds of regimes that do not satisfy his principles of justice—laissez-faire capitalism, welfare-state capitalism, state socialism, and liberal (democratic) socialism—in John Rawls, *Justice as Fairness: A Restatement* (Cambridge, MA: Harvard University Press, 2001), 135ff. In presenting some basic aspects of Rawls's theory, we will draw mostly on his 2001 *Restatement* but will ignore his "political liberalism"—a stance that abandons the moral universality he previously claimed for the ideals of free and equal citizenship and principles of justice he has defended. The *constructivist eudaimonism* we outline in the next section adopts aspects of Rawls's theory but is unabashedly morally universalistic, hence not a form of *political* liberalism but a form of *moral* liberalism. Readers new to these debates should also note that what political philosophers understand by "liberalism" has everything to do with ideals of free and equal citizenship and nothing to do with "neoliberalism" or neoliberal economic policy. Nor should it be equated with liberalism as a menu option in American politics.

41. Ibid., 5–7.

42. Ibid., 32–35.

43. Ibid., 27, 89–94.

44. Ibid., 89.

45. Ibid., 90 (italics added).

46. For a related statement on the authoritative role of science in democratic deliberations, see Gutmann and Thompson, *Democracy and Disagreement*, esp. 14–15, 65.

47. Rawls, *Justice as Fairness*, 90, 91.

48. Allen Buchanan, "Political Liberalism and Social Epistemology," *Philosophy & Public Affairs* 32, no. 2 (2004).

49. See Kitcher, "Public Knowledge and Its Discontents"; and *Science in a Democratic Society* (Amherst, NY: Prometheus Books, 2011).

50. Rawls, *Justice as Fairness*, 50.

51. Ibid., 59.

52. Ibid., 59.

53. Ibid., 58–59.

54. Rawls, "Fairness to Goodness," in *John Rawls: Collected Papers*, ed. Samuel Freeman (Cambridge, MA: Harvard University Press, 1999), 271.

55. Rawls, *Justice as Fairness*, 85–89; Rawls, *A Theory of Justice*, rev. ed. (Cambridge, MA: Harvard University Press, 1999), 119 (italics added).

56. Rawls, *Justice as Fairness*, 42–43.

57. Ibid., 43–44.

58. Ibid., 139.

59. Ibid., 42.

60. Because Rawls identifies primary goods as "all-purpose means" that citizens "need as free and equal persons living a complete life," it is reasonable to ask whether these goods could be adapted to permit long-term comparisons of aggregate opportunity. This too seems doubtful, given the presence of offices, positions, income, and wealth on Rawls's list. One would need an independent measure of the qualities of offices and positions important to living well, and income and wealth would need to be inflation adjusted, but could not be inflation adjusted across contexts that may not share a meaningful common basket of goods and services. Even if such a basket of goods and services did exist, it is far from clear that it would be a

good underlying measure of opportunities to live well. For an introduction to critiques of Rawls's primary goods as an adequate "metric" of justice, see Sen, *Development as Freedom*. For scholarly assessments of the competing claims of Rawls's primary goods approach and Sen's capability approach, see Harry Brighouse and Ingrid Robeyns, eds., *Measuring Justice: Primary Goods and Capabilities* (Cambridge: Cambridge University Press, 2010).

61. Danielle Zwarthoed has commented that some Rawlsians have tried to elaborate an intergenerational DP, but it would take us too far afield to examine the possible merits of their proposals for conceptualizing sustainability. See Daniel Attas, "A Transgenerational Difference Principle," in *Intergenerational Justice*, ed. Axel Gosseries and Lukas H. Meyer (Oxford: Oxford University Press, 2009); and Frédéric Gaspart and Axel Gosseries, "Are Generational Savings Unjust?," *Politics, Philosophy & Economics* 6, no. 2 (2007): 193–217.

62. Rawls, *Justice as Fairness*, 159–160.

63. Ibid., 159.

64. Ibid., 160.

65. See Richard Layard, *Lessons from a New Science* (London: Penguin, 2005), 33. Layard, a British economist, found in his study of an array of countries around the world that national average happiness scores stopped rising when a per capita income threshold of about $28,500 in US 2015 inflation-adjusted dollars was reached.

66. Rawls refers to the value of nature for future generations in *Political Liberalism* (New York: Columbia University Press, 1993), 245, noting that in a political liberal society, the preservation of the natural capital may be justified on the following grounds: "to further the good of ourselves and future generations by preserving the natural order and its life-sustaining properties; to foster species of animals and plants for the sake of biological and medical knowledge with its potential applications to human health; to protect the beauties of nature for purposes of public recreation and the pleasures of a deeper understanding of the world." However, he does not say this should be balanced with justice-based considerations that favor non-natural capital accumulation. We are grateful to Danielle Zwarthoed for her observations about this passage.

67. Rawls, *The Law of Peoples* (Cambridge, MA: Harvard University Press, 1999).

68. See Darrell Moellendorf, *Cosmopolitan Justice* (Boulder, CO: Westview Press, 2002); Martha Nussbaum, *Frontiers of Justice* (Cambridge, MA: Harvard University Press, 2006); and Gillian Brock, *Global Justice: A Cosmopolitan Account* (Oxford: Oxford University Press, 2009).

69. For a systematic overview of the theory of basic psychological needs, the larger theory of motivation, action, and well-being of which it is a component, and four

decades of supporting research, see Edward L. Deci and Richard M. Ryan, "Motiva-tion, Personality, and Development within Embedded Social Contexts: An Overview of Self-Determination Theory," in *The Oxford Handbook of Human Motivation*, ed. Richard Ryan (New York: Oxford University Press, 2012). See also Valery I. Chirkov, Richard M. Ryan, and Kennon M. Sheldon, eds., *Human Autonomy in Cross-Cultural Context: Perspectives on the Psychology of Agency, Freedom, and Well-Being* (Dordrecht, Netherlands: Springer, 2011).

70. See Ryan, Curren, and Deci, "What Humans Need." Psychologists do not use the word *admirable* and generally take pains to avoid value judgments, but—as we explain ahead—there are forms of objective goodness reliably associated with the satisfaction of psychological needs and fulfillment of related potentials.

71. Edward L. Deci et al., "On the Benefits of Giving as well as Receiving Autonomy Support: Mutuality in Close Friendships," *Personality and Social Psychology Bulletin* 32, no. 3 (2006).

72. See Chirkov, Ryan, and Sheldon, *Human Autonomy in Cross-Cultural Context.*

73. Franz De Wall, *The Age of Empathy: Nature's Lessons for a Kinder Society* (New York: Random House, 2009); and Lynn Stout, *Cultivating Conscience: How Good Laws Make Good People* (Princeton, NJ: Princeton University Press, 2011), pt. 2.

74. Stout, *Cultivating Conscience*, 92.

75. Ryan, Curren, and Deci, "What Humans Need," 67, summarizing the findings reported in Netta Weinstein and Richard M. Ryan, "When Helping Helps: Autono-mous Motivation for Prosocial Behavior and Its Influence on Well-Being for the Helper and Recipient," *Journal of Personality and Social Psychology* 98 (2010).

76. Layard, *Lessons from a New Science.*

77. Kasser, *High Price of Materialism*; Richard M. Ryan et al., "All Goals Are Not Cre-ated Equal: An Organismic Perspective on the Nature of Goals and Their Regula-tion," in *The Psychology of Action: Linking Cognition and Motivation to Behavior*, ed. Peter M. Gollwitzer and John A. Bargh (New York: Guilford, 1996); Tim Kasser and Richard M. Ryan, "Further Examining the American Dream: Differential Correlates of Intrinsic and Extrinsic Goals," *Personality and Social Psychology Bulletin* 22 (1996); and Christopher P. Niemiec, Richard M. Ryan, and Edward L. Deci, "The Path Taken: Consequences of Attaining Intrinsic and Extrinsic Aspirations in Post-College Life," *Journal of Research in Personality* 43 (2009).

78. See Kasser, *High Price of Materialism*, for a survey of much of the relevant research.

79. Readers familiar with the *capability approach* presented by the economist and philosopher Amartya Sen and philosopher Martha Nussbaum should note that the view presented here is similarly a "midfare" measure of justice, but we find it

analytically clearer to treat capabilities as one of three *internal* or personal necessities for living well and to distinguish such internal necessities from *external* necessities. The capability approach has been presented in different ways in a great number of works, but it is contrasted most systematically with Rawls's theory in Nussbaum's *Frontiers of Justice.* For a critique of foundational aspects of Nussbaum's version and a fuller defense of the alternative presented here, see Randall Curren, "Aristotelian Necessities," *Good Society—PEGS* 22, no. 2 (Fall 2013).

80. Buchanan, "Political Liberalism and Social Epistemology"; Curren, *Aristotle on the Necessity of Public Education.*

81. There are complexities that we can only touch on by identifying our assumptions: (1) MC and HC generally function in conjunction with NC, rather than replacing it (Costanza and Daly, "Natural Capital and Sustainable Development," 41); (2) MC tends to depreciate and must be replaced largely out of TNC; and (3) uncompensated depletion of NNC would be predictive of declining opportunity at some future time. It follows from the first and second of these assumptions that economic growth would not provide reliable compensation for loss of natural capital. Projections of economic expansion into the future would be predicated on an endless and most improbable expansion of ecological capacity or ecological debt, and without such continuing expansion MC and other economic assets would be subject to collapsing value. The MC would become obsolete without the NC on which it relies.

82. For an explanation of how a relevantly similar approach would work, see Paul Baer et al., "Greenhouse Development Rights: A Framework for Climate Protection That Is 'More Fair' than Equal Per Capita Emissions Rights," in *Climate Ethics*, ed. Stephen M. Gardiner et al. (Oxford: Oxford University Press, 2010).

83. See Maude Barlow, *Blue Future: Protecting Water for People and the Planet Forever* (New York: New Press, 2013).

4   Complexity and the Structure of Opportunity

1. For an account of *veritistic* institutions and developments of different parts of this suggestion, see Alvin Goldman, *Knowledge in a Social World* (New York: Oxford University Press, 1999); Buchanan, "Political Liberalism and Social Epistemology"; Kitcher, "Public Knowledge and Its Discontents"; Kitcher, *Science in a Democratic Society.*

2. Rawls, *Justice as Fairness*, 8–9.

3. See Margaret Heffernan, *Willful Blindness: Why We Ignore the Obvious* (New York: Walker and Co., 2011).

4. Kitcher, "Public Knowledge and Its Discontents," 109–110.

5. Ibid., 110.

6. We emphasize the word *legislative* because decisions about what science to publicly sponsor and permit would be periodic and time sensitive, not constitutional and enduring.

7. On the obstacles, see Katherine L. Jacobs, Gregg M. Garfin, and M. Lenart, "More than Just Talk: Connecting Science and Decision Making," *Environment* 47, no. 9 (2005).

8. Sandra S. Batie, "Wicked Problems and Applied Economics," *American Journal of Agricultural Economics* 90, no. 5 (2008): 1183. See also David H. Guston, "Boundary Organizations in Environmental Policy and Science," *Science, Technology & Human Values* 26, no. 4 (2001).

9. Batie, "Wicked Problems and Applied Economics," 1183.

10. Edward L. Deci et al., "Facilitating Internalization: The Self-Determination Theory Perspective," *Journal of Personality* 62, no. 1 (1994); Christopher P. Niemiec et al., "The Antecedents and Consequences of Autonomous Self-Regulation for College: A Self-Determination Theory Perspective on Socialization," *Journal of Adolescence* 29 (2006); and Deci and Ryan, "Motivation, Personality, and Development within Embedded Social Contexts."

11. Richard M. Ryan, "Psychological Needs and the Facilitation of Integrative Processes," *Journal of Personality* 63 (1995); and Kennon M. Sheldon and Tim Kasser, "Coherence and Congruence: Two Aspects of Personality Integration," *Journal of Personality and Social Psychology* 68, no. 3 (1995).

12. Deci et al., "Facilitating Internalization."

13. Ibid., 124.

14. Deci and Ryan, "Motivation, Personality, and Development within Embedded Social Contexts," 86, 93.

15. On the idea of learning as initiation pertaining to forms of goodness and not just skills or knowledge, see R. S. Peters, "Education as Initiation," in *Philosophy of Education: An Anthology*, ed. Randall Curren (Oxford: Blackwell, 2007); and Kenneth Strike, "The Ethics of Teaching," in *A Companion to the Philosophy of Education*, ed. Randall Curren (Oxford: Blackwell, 2003).

16. Curren, "Judgment and the Aims of Education."

17. The July 2009 issue of *Theory and Research in Education* (vol. 7, no. 2) provides a comprehensive overview of the relevant research up to the date of its publication.

18. See Kenneth Strike, *Small Schools & Strong Communities: A Third Way of School Reform* (New York: Teachers College Press, 2010).

19. Julia Annas, *Intelligent Virtue* (Oxford: Oxford University Press, 2011); Curren, "Motivational Aspects of Moral Learning and Progress"; Curren, "Virtue Ethics and Moral Education."

20. Frederick Herzberg, *Work and the Nature of Man* (New York: World Publishing, 1966).

21. Howard Gardner, Mihaly Csikszentmihalyi, and William Damon, *Good Work* (New York: Basic Books, 2001), 50.

22. Maarten Vansteenkiste et al., "On the Relations among Work Value Orientations, Psychological Need Satisfaction and Job Outcomes: A Self-Determination Theory Approach," *Journal of Occupational and Organizational Psychology* 80 (2007); Kasser et al., "Some Costs of American Corporate Capitalism"; Schor, *Plenitude*, 176ff.

23. See Schor, *Plenitude*, chap. 5, for a detailed discussion of better distributing employment through shorter work hours as a sustainable alternative to economic growth.

24. Tainter, *Collapse of Complex Societies*, 193, 197–199, 209–216. Tainter's other major works include Roderick J. McIntosh, Joseph A. Tainter, and Susan Keech McIntosh, eds., *The Way the Wind Blows: Climate, History, and Human Action* (New York: Columbia University Press, 2000); T. F. H. Allen, Joseph A. Tainter, and Thomas W. Hoekstra, *Supply-Side Sustainability* (New York: Columbia University Press, 2003); and Tainter and Patzek, *Drilling Down*.

25. See chapter 3.

26. For a sampling, see Tainter, *Collapse of Complex Societies*, chap. 3; Charles Redman, *Human Impact on Ancient Environments*; Ronald Wright, *A Short History of Progress* (Toronto: House of Anansi Press, 2004); Diamond, *Collapse*; McAnany and Yoffee, *Questioning Collapse*; Robert Costanza, Lisa J. Graumlich, and Will Steffen, eds., *Sustainability or Collapse? An Integrated History and Future of People on Earth* (Cambridge, MA: MIT Press, 2011).

27. Tainter, *Collapse of Complex Societies*, 4.n

28. Ibid., 37, 93.

29. Ibid., 119.

30. Ibid., chap. 4, "Understanding Collapse: The Marginal Productivity of Sociopolitical Change." Tainter and Patzek, *Drilling Down*, 84–134, covers some of the same ground and adds valuable, telling detail with respect to technological innovation, efficiency, and declining return on investment.

31. Schor, *Born to Buy*; and Schor, *The Overworked American: The Unexpected Decline of Leisure* (New York: Basic Books, 1992). See also Schor, *Plenitude*, on the logic of fashion and dramatic acceleration of consumption in recent years.

32. Tainter, *Collapse of Complex Societies*, 201.

33. See, esp., Joseph Tainter, T. F. H. Allen, and Thomas W. Hoekstra, "Energy Transformations and Post-Normal Science," *Energy* 31 (2006); and Tainter and Patzek, *Drilling Down*.

34. Tainter and his coauthor, Tadeusz Patzek, touch upon the challenges of managing the complexities of a human and mechanical system as complex as a deep water drilling rig in their book on the BP oil spill, *Drilling Down*, but they do not make the connection to the costs of corporate governance more generally. We will address the BP oil spill as one of three case studies in the next chapter.

35. Horace Mann, "Twelfth Annual Report," in *The Republic and the School: Horace Mann on the Education of Free Men*, ed. Lawrence Cremin (New York: Teachers College Press, 1957), 85, 87; quoted in David Labaree, *How to Succeed in School without Really Learning: The Credentials Race in American Education* (New Haven, CT: Yale University Press, 1997), 20–21.

36. See Thomas F. Green, *Predicting the Behavior of the Educational System* (Syracuse, NY: Syracuse University Press, 1980). Green dates the identification of dropping out as a social problem to the mid-1950s, as high school completion reached about 60 percent.

37. Thomas Green introduced the term *group of last entry* in *Predicting the Behavior of the Educational System*. Although the abstract model presented in this work was grounded in multinational data tracking the growth of educational systems across a century of expansion, like other accounts of that era it did not consider particularities of national contexts and histories that are important to explaining the way the US system functions. Cf. Raymond Boudon, *Education, Opportunity, and Social Inequality: Changing Prospects in Western Society* (New York: Wiley, 1974); Ronald Dore, *The Diploma Disease: Education, Qualification, and Development* (London: Allen and Unwin, 1976); and Randall Collins, *The Credential Society: An Historical Sociology of Education and Stratification* (New York: Academic Press, 1979).

38. For historical accounts of how this came about, see David K. Brown, *Degrees of Control: A Sociology of Educational Expansion and Occupational Credentialism* (New York: Teachers College Press, 1995); and Hal Hansen, "Rethinking Certification Theory and the Educational Development of the United States and Germany," *Research in Social Stratification and Mobility* 29 (2011). Public high schools in the United States were in many areas initially regarded as "people's colleges" and they competed directly with colleges and proprietary technical and professional schools. It was only by 1920 and through some historical contingencies that the University of Michigan's decision to offer college admission to all the graduates of high schools that met its stipulated accreditation standards proved to be a decisive turning point in the embedding of high schools within an integrated hierarchical system.

39. The term *contest mobility* was introduced by Ralph Turner in "Sponsored and Contest Mobility and the School System," *American Sociological Review* 25 (1960). Influential functionalist theories of schools as mere replicators of privilege include Samuel Bowles and Herbert Gintis, *Schooling in Capitalist America* (New York: Basic Books, 1976); and Pierre Bourdieu and Jean-Claude Passeron, *Reproduction in Education, Society, and Culture* (Beverly Hills, CA: Sage Publications, 1977). For a critique of such theories, see David P. Baker, "Forward and Backward, Horizontal and Vertical: Transformation of Occupational Credentialing in the Schooled Society," *Research in Social Stratification and Mobility* 29 (2011).

40. Rawls may have assumed that children could be reliably sorted by differences of native talent and streamed into different curricula without social class bias. In that case, FEO could be achieved without preparing every child for college. It is now clear that the tests that have been used to sort and track students do not measure a fixed capacity to learn and that intellectual "gifts" or talents more generally are largely a product of the quantity and quality of early learning. Sorting students by talent cohorts, in ways that amplify the differences discerned at an early age, has thus fallen out of favor as a model of equal opportunity. Thus, with college now a virtual necessity for middle-class status, making social class origins less predictive of who goes on to (and finishes) college is a major focus of educational practice and theory. EO seems to require that society provide students with ways to catch up, at least through college, and for this reason and others, the politics of higher education are focused on expanding access to college education. Rawls's concept of *fair equality of opportunity* has been interpreted in a way that is consistent with this understanding of EO. It invokes the very idea of native talents that has fallen out of favor, but this is implicitly corrected by neo-Rawlsians, who assume that the focus of egalitarian educational reform should be on college admission and completion.

41. Labaree, *How to Succeed in School Without Really Learning*, 50.

42. David Bills, *Sociology of Education and Work* (Malden, MA: Blackwell, 2004); David Baker, "The Educational Transformation of Work: Towards a New Synthesis," *Journal of Education and Work* 22 (2009); and David Baker, *The Schooled Society: The Educational Transformation of Global Culture* (Stanford, CA: Stanford University Press, 2014).

43. Cf. Harry Brighouse's relatively mild view that higher education confers further (unjust) advantages on those who are already advantaged by accident of birth; Brighouse, "Globalization and the Professional Ethic of the Professoriat," in *Global Inequalities and Higher Education*, ed. Elaine Unterhalter and Vincent Carpentier (New York: Palgrave Macmillan, 2010).

44. Tainter, *Collapse of Complex Societies*, 91–126.

45. David Labaree details some of the reasons for this in *Someone Has to Fail: The Zero-Sum Game of Public Schooling* (Cambridge, MA: Harvard University Press, 2010).

46. Duane Elgin, *Voluntary Simplicity*, 2nd rev. ed. (New York: Harper, 2010); Cecile Andrews, *The Circle of Simplicity* (New York: Harper, 1997).

47. Xiaoling Shu and Margaret Mooney Marini, "Coming of Age in Changing Times: Occupational Aspirations of American Youth, 1966–1980," *Research in Social Stratification and Mobility* 26, no. 1 (2008); and Hansen, "Rethinking Certification Theory."

48. Labaree, *How to Succeed in School without Really Learning*, 35; L. Pelletier and E. Sharp, "Administrative Pressures and Teachers' Interpersonal Behavior," *Theory and Research in Education* 7, no. 2 (2009); Richard M. Ryan and Netta Weinstein, "Undermining Quality Teaching and Learning: A Self-Determination Theory Perspective on High-Stakes Testing," *Theory and Research in Education* 7, no. 2 (2009); and Maarten Vansteenkiste, Bart Soenens, Joke Verstuyf, and Willy Lens, "'What is The Usefulness of Your Schoolwork?' The Differential Effects of Intrinsic and Extrinsic Goal Framing on Optimal Learning," *Theory and Research in Education* 7, no. 2 (2009).

49. Tainter notes that regulation and resistance to it can generate "an unending spiral ... with complexity and costs continuously increasing" (*Collapse of Complex Societies*, 116), but he does not dwell on what this implies.

50. Stanley Katz, "Choosing Justice over Excellence," *Chronicle of Higher Education* 48, no. 35 (May 17, 2002), http://www.princeton.edu/~snkatz/papers/CHE_justice .html; and Katz, "The Pathbreaking, Fractionalized, Uncertain World of Knowledge," *Chronicle of Higher Education* 49, no. 4 (September 20, 2002), http://www.princeton .edu/~snkatz/papers/CHE_knowledge.html.

51. See, e.g., Jennifer Everett, "Sustainability in Higher Education: Implications for the Disciplines," *Theory and Research in Education* 6, no. 2 (2008).

52. Hansen, "Rethinking Certification Theory," 48–49. For more on the German system and the role of training more generally, see Alison Wolf, *Does Education Matter? Myths about Education and Economic Growth* (London: Penguin, 2002).

53. Hansen, "Rethinking Certification Theory," 34.

54. Ibid.

55. Ibid., 50; citing Matthias Pilz, "Why *Abiturienten* Do an Apprenticeship before Going to University," *Oxford Review of Education* 35, no. 2 (2009).

56. Christian Smith, *Lost in Transition: The Dark Side of Emerging Adulthood* (New York: Oxford University Press, 2011); and Hansen, "Rethinking Certification Theory."

57. See Joel Feinberg, "The Child's Right to an Open Future," in *Philosophy of Education: An Anthology*, ed. Randall Curren (Oxford: Blackwell, 2007).

58. Brian Schrag, "Moral Responsibility of Faculty and the Ethics of Faculty Governance," in *Ethics in Academia*, ed. S. K. Majumdar, Howard S. Pitkow, Lewis Penhall Bird, and E. W. Miller (Easton: Pennsylvania Academy of Science, 2000), 232; and Peter Markie, *A Professor's Duties: Ethical Issues in College Teaching* (Lanham, MD: Rowman & Littlefield, 1994), 16.

59. Schrag, "Moral Responsibility of Faculty and the Ethics of Faculty Governance," 234.

60. See, e.g., Engineering Council UK, *Guidance on Sustainability for the Engineering Profession* (London: Engineering Council UK, 2009), http://www.engc.org.uk/engcdocuments/internet/Website/Guidance%20on%20Sustainability.pdf. The US National Society of Professional Engineers Code of Ethics includes the following provision: "Engineers are encouraged to adhere to the principles of sustainable development in order to protect the environment for future generations." See http://www.nspe.org/resources/ethics/code-ethics.

61. UNESCO, Education for Sustainable Development: United Nations Decade (2005–2014) (UNESCO, 2005), http://en.unesco.org/themes/education-sustainable-development.

62. Schor, *Plenitude*.

5   Managing Complexity: Three Case Studies

1. On nonideal theory, see David Miller, "Political Philosophy for Earthlings," in *Political Theory: Methods and Approaches*, ed. David Leopold and Marc Stears (Oxford: Oxford University Press, 2008); A. John Simmons, "Ideal and Non-Ideal Theory," *Philosophy and Public Affairs* 38, no. 1 (2010); and Zofia Stemplowska and Adam Swift, "Ideal and Non-Ideal Theory," in *The Oxford Handbook of Political Philosophy*, ed. David Estlund (New York: Oxford University Press, 2012). On wicked problems, see Horst W. J. Rittel and Melvin M. Webber, "Dilemmas in a General Theory of Planning," *Policy Sciences* 4, no. 2 (1973); Batie, "Wicked Problems and Applied Economics"; Jeff Conklin, *Wicked Problems and Social Complexity* (Napa, CA: CogNexus Institute, 2006); and Catrien J. A. M. Termeer et al., "Governance Capabilities for Dealing Wisely with Wicked Problems," *Administration & Society* 47, no. 6 (2015).

2. Cf. H. C. Peterson, "Sustainability: A Wicked Problem," in *Sustainable Animal Agriculture*, ed. Ermias Kebreab (Wallingford, UK: CABI, 2013). For other works on sustainability in which the idea of a wicked problem plays a major role, see Bryan Norton, *Sustainable Values, Sustainable Change* (Chicago: University of Chicago Press, 2015); and Jenkins, *Future of Ethics*.

3. See Moira Zellner and Scott D. Campbell, "Planning for Deep-Rooted Problems: What Can We Learn from Aligning Complex Systems?," *Planning Theory & Practice* 16, no. 4 (2015).

4. Rittel and Webber, "Dilemmas in a General Theory of Planning," 160.

5. Batie, "Wicked Problems and Applied Economics," 1176.

6. Richard R. Nelson, "Intellectualizing about the Moon-Ghetto Metaphor: A Study of the Current Malaise of Rational Analysis of Social Problems," *Policy Sciences* 5, no. 4 (1974): 376.

7. Thaddeus R. Miller, *Reconstructing Sustainability Science* (London: Routledge, 2014), 7.

8. Batie, "Wicked Problems and Applied Economics."

9. Brian W. Head, "Wicked Problems in Public Policy," *Public Policy* 3, no. 2 (2008).

10. John D. Sterman, "Learning from Evidence in a Complex World," *American Journal of Public Health* 96, no. 3 (2006): 3.

11. Tainter and Patek, *Drilling Down*, 75.

12. Ibid.

13. Dryzek, *Rational Ecology*; Kelly Levin et al., "Overcoming the Tragedy of Super Wicked Problems: Constraining Our Future Selves to Ameliorate Global Climate Change," *Policy Sciences* 45, no. 2 (2012); and Ross Garnaut, *The Garnaut Climate Change Review* (Cambridge: Cambridge University Press, 2008).

14. Much the same is true of the problems we face in not breaching other planetary boundaries.

15. There is something to be said for avoiding the suggestions of evil and discouragement that run through many discussions of "wicked problems." Consider the reference to *malaise* in Nelson, "Intellectualizing about the Moon-Ghetto Metaphor." The frustration and apparent discouragement in Rittel and Webber's characterization of the wickedness that planners endure is palpable; there are no "definitive," "true-or-false," uncontested, or objectively tested solutions, only ones "good-or-bad in the eyes of stakeholders"; "there is no public tolerance of initiatives or experiments that fail," and there are no second chances ("Dilemmas in a General Theory of Planning," 155, 162, 163, 166). Consider finally the gratuitous intimation of evil in Garnaut, *Garnaut Climate Change Review*: "It [climate change] is insidious rather than (as yet) directly confrontational" (xviii).

16. Batie, "Wicked Problems and Applied Economics."

17. Australian Public Service Commission, *Tackling Wicked Problems: A Public Policy Perspective* (Canberra: Australian Public Service Commission, 2007), iii.

18. Brian W. Head and John Alford, "Wicked Problems: Implications for Public Policy and Management," *Administration & Society* 47, no. 6 (2015); Norton, *Sustainable Values, Sustainable Change*.

19. Jeremy Allouche, Carl Middleton, and Dipak Gyawali, "Nexus Nirvana or Nexus Nullity? A Dynamic Approach to Security and Sustainability in the Water-Energy-Food Nexus," STEPS Working Paper 63, STEPS Centre, Brighton, UK, 2014.

20. Columbia Accident Investigation Board, *Report of the Columbia Accident Investigation Board* (Washington, DC: National Aeronautics and Space Administration, 2003).

21. Thomas Princen, Jack P. Manno, and Pamela Martin, "Keep Them in the Ground: Ending the Fossil Fuel Era," in *State of the World 2013*, ed. Worldwatch Institute (Washington, DC: Island Press, 2013).

22. Charles A. S. Hall, Jessica G. Lambert, and Stephen B. Balogh, "EROI of Different Fuels and the Implications for Society," *Energy Policy* 64 (2014).

23. Tainter and Patzek, *Drilling Down*, 200.

24. David J. Murphy, "The Implications of the Declining Energy Return on Investment of Oil Production," *Philosophical Transactions of the Royal Society of London A: Mathematical, Physical and Engineering Sciences* 372, no. 2006 (2014), doi:10.1098/rsta.2013.0126, http://rsta.royalsocietypublishing.org/content/372/2006/20130126.

25. Tainter and Patzek, *Drilling Down*, 210.

26. Ian Urbina and Justin Gillis, "Workers on Oil Rig Recall a Terrible Night of Blasts," *New York Times*, May 7, 2010, http://www.nytimes.com/2010/05/08/us/08rig.html?pagewanted=all. Except where noted, the account that follows is based on Bob Graham et al., *Deep Water: The Gulf Oil Disaster and the Future of Offshore Drilling* (Washington, DC: United States Publishing Office, 2011).

27. Exploratory wells like the one drilled by Deepwater Horizon seek oil and gas in deeply buried rock layers that are under enormous pressure due to the weight of the overlying rock; the deeper the well, the higher the pressure. The biggest challenge of deep-water drilling is thus controlling pressure to prevent a *blowout* (i.e., uncontrolled flow of hydrocarbons into the well). Well control typically involves applications of two or more independently acting barriers. Drilling fluid or "mud" is injected into the well to counterbalance the pressure arising from hydrocarbon-rich rock layers. Strings of casing pipe and strategically placed cement plugs provide additional barriers. The last line of defense in the battle to control pressure is a gigantic assembly of valves known as the *blowout preventer* (BOP). See Deepwater Horizon Study Group, *Final Report on the Investigation of the Macondo Well Blowout*, March 1, 2011, http://www.aspresolver.com/aspresolver.asp?ENGV;2082362.

28. Russell Gold and Ben Casselman, "On Doomed Rig's Last Day, a Divisive Change of Plan," *Wall Street Journal*, August 26, 2010, http://online.wsj.com/article/.

29. Cutler J. Cleveland, C. Michael Hogan, and Peter Saundry, "Deepwater Horizon Oil Spill," in *Encyclopedia of Earth*, ed. Cutler J. Cleveland (Washington, DC:

Environmental Information Coalition, National Council for Science and the Environment, 2010), http://www.eoearth.org/view/article/161185/.

30. "Timeline: Oil Spill in the Gulf," CNN.com, n.d., www.cnn.com/2010/US/05/03/timeline.gulf.spill/index.html.

31. Cleveland, "Deepwater Horizon Oil Spill."

32. Daniel A. Farber, "The BP Blowout and the Social and Environmental Erosion of the Louisiana Coast," *Minnesota Journal of Law, Science & Technology* 13 (2012); Jason Mark, "We Are All Louisianans," *Earth Island Journal* 25, no. 3 (Autumn 2010), http://www.earthisland.org/journal/index.php/eij/article/we_are_all_louisianans; and Debbie Elliott, "Five Years after BP Oil Spill, Effects Linger and Recovery Is Slow," *NPR*, April 20, 2015, http://www.npr.org/2015/04/20/400374744/5-years-after-bp-oil-spill-effects-linger-and-recovery-is-slow.

33. Farber, "The BP Blowout and the Social and Environmental Erosion of the Louisiana Coast."

34. Elliott, "Five Years after BP Oil Spill."

35. Farber, "The BP Blowout and the Social and Environmental Erosion of the Louisiana Coast."

36. Warren Cornwall, "Deepwater Horizon: After the Oil." *Science* 348, no. 6230 (2015).

37. Mark A. Latham, "Five Thousand Feet and Below: The Failure to Adequately Regulate Deepwater Oil Production Technology," *Boston College Environmental Affairs Law Review* 38, no. 2 (2011), 347, http://lawdigitalcommons.bc.edu/cgi/viewcontent.cgi?article=1692&context=ealr.

38. John McQuaid, "The Gulf of Mexico Oil Spill: An Accident Waiting to Happen," *Yale Environment* 360 (2010), http://e360.yale.edu/feature/the_gulf_of_mexico_oil_spill_an_accident_waiting_to_happen/2272/.

39. John McQuaid, "Gulf of Mexico Oil Spill."

40. "BP's Troubled Past," *PBS Frontline*, October 26, 2010, http://www.pbs.org/wgbh/pages/frontline/the-spill/bp-troubled-past/.

41. Heffernan, *Willful Blindness*.

42. "BP's Troubled Past."

43. Heffernan, *Willful Blindness*.

44. Ibid.

45. John McQuaid, "Gulf of Mexico Oil Spill"; and Raymond Wassel, "Lessons from the Macondo Well Blowout in the Gulf of Mexico," *Bridge* 44, no. 3 (2014).

46. Elaine M. Brown, "The Deepwater Horizon Disaster," in *Case Studies in Organizational Communication: Ethical Perspectives and Practices*, ed. Steve May (Thousand Oaks, CA: Sage Press, 2012).

47. Graham et al., *Deep Water*.

48. Ibid., 133.

49. Robert Bea, "Understanding the Macondo Well Failures," Deepwater Horizon Study Group Working Paper, January 2011, http://ccrm.berkeley.edu/pdfs_papers/DHSGWorkingPapersFeb16-2011/UnderstandingMacondoWellFailures-BB_DHSG-Jan2011.pdf.

50. Anthony Mays, ed., *Disaster Management: Enabling Resilience* (Dordrecht, Netherlands: Springer, 2015), v.

51. Venkat Venkatasubramanian, "Systemic Failures: Challenges and Opportunities in Risk Management in Complex Systems," *AIChE Journal* 57, no. 1 (2011), 2–3.

52. Associated Press, "5 Years after BP Spill, Drillers Push into Riskier Depths," *Chicago Tribune*, April 20, 2015, http://www.chicagotribune.com/news/nationworld/chi-bp-oil-spill-20150419-story.html; and Coral Davenport, "U.S. Will Allow Drilling for Oil in Arctic Ocean," *New York Times*, May 11, 2015, http://www.nytimes.com/2015/05/12/us/white-house-gives-conditional-approval-for-shell-to-drill-in-arctic.html.

53. IPCC, *Climate Change 2014: Synthesis Report*, ed. Core Writing Team, Rajendra K. Pachuri, and Leo Meyer (Geneva: IPCC, 2014), https://www.ipcc.ch/report/ar5/syr/.

54. Andrea Thompson, "CO2 Nears Peak: Are We Permanently above 400 PPM?," *Climate Central*, May 16, 2016, http://www.climatecentral.org/news/co2-are-we-permanently-above-400-ppm-20351.

55. International Energy Agency, "World Energy Outlook 2014 Factsheet" (International Energy Agency, 2015), http://www.worldenergyoutlook.org/media/weowebsite/2014/141112_WEO_FactSheets.pdf.

56. WWF, *Living Planet Report 2014*, 91.

57. Heather Cooley et al., *Global Water Governance in the 21st Century* (Oakland, CA: Pacific Institute, 2013); and Guy Pegram et al., *River Basin Planning Principles: Procedures and Approaches for Strategic Basin Planning* (Paris: UNESCO, 2013).

58. R. Quentin Grafton et al., "Global Insights into Water Resources, Climate Change and Governance," *Nature Climate Change* 3, no. 4 (2013).

59. Cooley et al., *Global Water Governance in the 21st Century*.

60. Pegram et al., *River Basin Planning Principles*; Helle Munk Ravnborg and Maria del Pilar Guerrero, "Collective Action in Watershed Management—Experiences from the Andean Hillsides," *Agriculture and Human Values* 16, no. 3 (September 1999).

61. National Water Commission, *Water Markets in Australia: A Short History* (Canberra: National Water Commission, 2011).

62. Nicholas Breyfogle, "Dry Days Down Under: Australia and the World Water Crisis," *Origins* 3, no. 7 (2010), http://origins.osu.edu/article/dry-days-down-under -australia-and-world-water-crisis; and Erin Musiol, Nija Fountano, and Andreas Safakas, "Drought Planning in Practice," in *Planning and Drought*, ed. James C. Schwab (Chicago: American Planning Association, 2013).

63. Breyfogle, "Dry Days Down Under."

64. Ibid.

65. R. Quentin Grafton et al., "Water Planning and Hydro-Climatic Change in the Murray-Darling Basin, Australia," *Ambio* 43, no. 8 (2014).

66. Amy Sennett et al., "Challenges and Responses in the Murray-Darling Basin," *Water Policy* 16, no. S1 (2014), doi:10.2166/wp.2014.006, http://wp.iwaponline. com/content/16/S1/117; Musiol, Fountano, and Safakas, "Drought Planning in Practice."

67. Grafton et al., "Water Planning and Hydro-Climatic Change in the Murray-Darling Basin, Australia."

68. Pegram et al., *River Basin Planning*.

69. Sennett et al., "Challenges and Responses in the Murray-Darling Basin."

70. NSW Government, "Algal Information," New South Wales Department of Primary Industries: Water, n.d., http://www.water.nsw.gov.au/Water-Management/ Water-quality/Algal-information/Dangers-and-problems/Dangers-and-problems/ default.aspx.

71. Sennett et al., "Challenges and Responses in the Murray-Darling Basin."

72. Adrian Piani, "The Key Ingredients for Success of the Murray-Darling Basin Plan" (Global Water Forum, 2013), http://www.globalwaterforum.org/2013/04/01/ the-key-ingredients-for-successof-the-murray-darling-basin-plan/.

73. Piani, "Key Ingredients for Success of the Murray-Darling Basin Plan."

74. Musiol et al., "Drought Planning in Practice"; Piani, "Murray-Darling Basin Background Paper."

75. Piani, "Key Ingredients for Success of the Murray-Darling Basin Plan."

76. Grafton et al., "Global Insights into Water Resources, Climate Change and Governance."

77. Amir AghaKouchak et al., "Australia's Drought: Lessons for California," *Science* 343, no. 6178 (2014).

78. Piani, "Key Ingredients for Success of the Murray-Darling Basin Plan."

79. Musiol et al., "Drought Planning in Practice."

80. National Water Commission, *Water Markets in Australia*.

81. National Water Commission, *Australia's Water Blueprint: National Reform Assessment 2014* (Canberra: National Water Commission, 2014), x.

82. Wentworth Group of Concerned Scientists, "Statement on the Future of Australia's Water Reform," Wentworth Group of Concerned Scientists, October 10, 2014, http://wentworthgroup.org/2014/10/statement-on-the-future-of-australias -water-reform/2014/.

83. Melanie Gale et al., "The Boomerang Effect: A Case Study of the Murray-Darling Basin Plan," *Australian Journal of Public Administration* 73, no. 2 (2014).

84. Ibid.

85. R. Quentin Grafton et al., "An Integrated Assessment of Water Markets: A Cross-country Comparison," *Review of Environmental Economics and Policy* 5, no. 2 (2011).

86. Shiney Varghese, "Water Governance in the 21st Century: Lessons from Water Trading in the U.S. and Australia" (IATP, 2013), http://www.iatp.org/documents/ water-governance-in-the-21st-century.

87. Barlow, *Blue Future*, 82.

88. Ibid. For further case studies and analyses, see Brown and Schmidt, *Water Ethics*.

89. Ian Campbell, Barry Hart, and Chris Barlow, "Integrated Management in Large River Basins: 12 Lessons from the Mekong and Murray-Darling Rivers," *River Systems* 20, no. 3–4 (2013).

90. Joyeeta Gupta, Claudia Pahl-Wostl, and Ruben Zondervan, "Global Water Governance: A Multi-level Challenge in the Anthropocene," *Current Opinion in Environmental Sustainability* 5, no. 6 (2013).

91. Ravnborg and Guerrero, "Collective Action in Watershed Management"; and Juan-Camilo Cardenas, Luz Angela Rodriguez, and Nancy Johnson, "Vertical Collective Action: Addressing Vertical Asymmetries in Watershed Management" (CEDE, February 2015), doi:10.13140/RG.2.1.2701.7767, https://www.researchgate.net/ publication/275351883_Vertical_Collective_Action_Addressing_Vertical _Asymmetries_in_Watershed_Management. See also the US Department of the

Interior, Bureau of Reclamation's Cooperative Watershed Management Program initiative (http://www.usbr.gov/watersmart/cwmp/index.html).

92. Nathan L. Engle et al., "Integrated and Adaptive Management of Water Resources: Tensions, Legacies, and the Next Best Thing," *Ecology and Society* 16, no. 1 (2011), http://www.ecologyandsociety.org/vol16/iss1/art19/.

93. Engle et al., "Integrated and Adaptive Management of Water Resources."

94. Philip J. Wallis and Raymond L. Ison, "Appreciating Institutional Complexity in Water Governance Dynamics: A Case from the Murray-Darling Basin, Australia," *Water Resources Management* 25, no. 15 (2011).

95. Ed Morgan, "Science in Sustainability: A Theoretical Framework for Understanding the Science-Policy Interface in Sustainable Water Resource Management," *International Journal of Sustainability Policy and Practice* 9 (2014).

96. Derek Armitage et al. "Science-Policy Processes for Transboundary Water Governance," *AMBIO* 44, no. 5 (2015): 353–366.

97. This is attributed to the French jurist and gastronome Jean Anthelme Brillat-Savarin (1755–1826). *Columbia World of Quotations* (New York: Columbia University Press, 1996).

98. Diamond, *Collapse*, 222–230.

99. Plutarch, "Lycurgus," XII.1–4, in *Plutarch's Lives*, vol. 1, trans. Bernadette Perrin (Cambridge, MA: Harvard University Press, 1914).

100. Will Hueston and Anni McLeod, "Overview of the Global Food System: Changes over Time/Space and Lessons for Future Food Safety," in *Improving Food Safety through a One Health Approach: Workshop Summary*, ed. Eileen R. Choffres et al. (Washington, DC: National Academies Press/Institute of Medicine, 2012), http://www.ncbi.nlm.nih.gov/books/NBK114491/.

101. Terry Marsden and Adrian Morley, "Current Food Questions and Their Scholarly Challenges," in *Sustainable Food Systems: Building a New Paradigm*, ed. Terry Marsden and Adrian Morley (New York: Routledge, 2014).

102. Sachs, *Age of Sustainable Development*, 338.

103. Marsden and Morley, *Sustainable Food Systems*.

104. Simon Bager, "Big Facts: Focus on East and Southeast Asia," CGIAR, May 6, 2014, http://ccafs.cgiar.org/blog/big-facts-focus-east-and-southeast-asia#.VQc6ZY7F98E.

105. Keskinen et al., "The Water-Energy-Food Nexus and the Transboundary Context: Insights from Large Asian Rivers," *Water* 8, no. 5 (2016).

106. Ibid.

107. David Fullbrook, "Food Security in the Wider Mekong Region," in *The Water-Food-Energy Nexus in the Mekong Region*, ed. Alexander Smajgl and John Ward (New York: Springer, 2013).

108. UNESCAP, *The Status of the Water-Food-Energy Nexus in Asia and the Pacific* (Bangkok: United Nations Economic and Social Commission for Asia and the Pacific, 2013), http://www.unescap.org/publications/detail.asp?id=1551.

109. Robyn M. Johnston et al., *Rethinking Agriculture in the Greater Mekong Subregion: How to Sustainably Meet Food Needs, Enhance Ecosystem Services and Cope with Climate Change* (Colombo, Sri Lanka: International Water Management Institute, 2010).

110. Johnston et al., *Rethinking Agriculture in the Greater Mekong Subregion*.

111. Jacqui Griffiths and Rebecca Lambert, *Free Flow: Reaching Water Security through Cooperation* (Geneva: UNESCO, 2013).

112. Griffiths and Lambert, *Free Flow*.

113. Fullbrook, "Food Security in the Wider Mekong Region."

114. Armitage et al., "Science-Policy Processes for Transboundary Water Governance."

115. UNESCAP, *Status of the Water-Food-Energy Nexus in Asia and the Pacific*.

116. Mekong River Commission, "Agreement on the Cooperation for the Sustainable Development of the Mekong River Basin," Mekong River Commission, April 5, 1995, http://www.mrcmekong.org/assets/Publications/policies/agreement-Apr95.pdf.

117. Ezra Ho, "Unsustainable Development in the Mekong: The Price of Hydropower," *Consilience: The Journal of Sustainable Development* 12, no. 1 (2014).

118. Jamie Pittock, "Devil's Bargain? Hydropower vs. Food Trade-Offs in the Mekong Basin," *World Rivers Review* 29, no. 4 (2014).

119. Kirk Herbertson, "Xayaburi Dam: How Laos Violated the 1995 Mekong Agreement," January 13, 2013, https://www.internationalrivers.org/blogs/267/xayaburi-dam-how-laos-violated-the-1995-mekong-agreement.

120. Ibid.

121. Christinia Larsen, "Mekong Megadrought Erodes Food Security," April 6, 2016, http://www.sciencemag.org/news/2016/04/mekong-mega-drought-erodes-food-security.

122. Ibid.

123. François Molle, Tira Foran, and Mira Kakonen, *Contested Waterscapes in the Mekong Region: Hydropower, Livelihoods and Governance* (London: Earthscan, 2012), 16.

124. Ibid.

6 An Education in Sustainability

1. Plato, *Laws*, VII 811c–d, in *Plato: Complete Works*, edited by John Cooper (Indianapolis: Hackett Publishing, 1997), 1478–1479. Plato's dialogues were written on papyrus rolls referred to as books, and the *Laws* filled ten books. A modern edition of the *Laws* has the appearance of a book in ten chapters.

2. The designation *education for sustainability* (EFS) is widely used in North America and Australia, sometimes interchangeably with *education for sustainable development* (ESD) to refer to DESD implementation. Mindy Spearman treats ESD and EFS as referring to the same (not well-defined) concept in "Sustainability Education," in *Educating about Social Issues in the 20th and 21st Centuries*, vol. 1, *A Critical Annotated Bibliography*, ed. Samuel Totten and Jon E. Pedersen (Charlotte, NC: Information Age Publishers, 2012). One of Spearman's data points centers on Andres Edwards, who describes sustainability and sustainability education in ways that make them indistinguishable from SD and ESD in *The Sustainability Revolution* (Gabriola Island, British Columbia: New Society Publishers, 2005), 20–23. Kate Sherren "takes the two alternate terms [EFS and ESD] as synonymous" in "A History of the Future of Higher Education for Sustainable Development," *Environmental Education Research* 14, no. 3 (2008): 238. A preference for EFS usually reflects concern about tensions between sustainability and development (noted in chapter 1); see Daniel Bonevac, "Is Sustainability Sustainable?," *Academic Questions* 23 (2010); and David Selby, "The Firm and Shaky Ground of Education for Sustainable Development," in *Green Frontiers: Environmental Educators Dancing Away from Mechanism*, ed. James Gray-Donald and David Selby (Rotterdam: Sense Publishers, 2008). We prefer education in sustainability (EiS) for reasons pertaining to matters of educational justification, authority, and aims.

3. WCED (World Commission on Environment and Development), *Our Common Future* (Geneva: United Nations, 1987), 12, http://www.un-documents.net/wced-ocf .htm.

4. UNESCO, Education for Sustainable Development: United Nations Decade (2005–2014) website (UNESCO, 2005), http://en.unesco.org/themes/education -sustainable-development, under Background.

5. IUCN, *Caring for the Earth: A Strategy for Sustainable Living* (Gland, Switzerland: IUCN, 1991).

6. Douglas Bourn, "Education for Sustainable Development and Global Citizenship: The UK Perspective," *Applied Environmental Education and Communication* 4, no. 3 (2005): 233.

7. UNESCO, Education for Sustainable Development: United Nations Decade (2005–2014) website (UNESCO, 2005), http://en.unesco.org/themes/education -sustainable-development.

8. Ibid., under Implementation.

9. Ibid., under Objectives and Strategies.

10. Ibid., under Quality Education.

11. Ibid.

12. Ibid.

13. UNESCO, Education for Sustainable Development: United Nations Decade (2005–2014) website (UNESCO, 2005), http://en.unesco.org/themes/education -sustainable-development.

14. UNESCO, *Guidelines and Recommendations for Reorienting Teacher Education to Address Sustainability* (Paris: UNESCO, 2005), http://unesdoc.unesco.org/images/ 0014/001433/143370E.pdf.

15. See, e.g., Department for Children, Schools and Families, *Brighter Futures— Greener Lives: Sustainable Development Action Plan 2008–2010* (2008), 15; Peter Higgins and Gordon Kirk, "Sustainability Education in Scotland: The Impact of National and International Initiatives on Teacher Education and Outdoor Education," *Journal of Geography in Higher Education* 30, no. 2 (2006); and Department of the Environment, Water, Heritage and the Arts, *Living Sustainably: The Australian Government's National Action Plan for Education for Sustainability* (Canberra: Department of the Environment, Water, Heritage and the Arts, 2009), 10, http://www.environment .gov.au/education/publications/pubs/national-action-plan.pdf.

16. Bourn, "Education for Sustainable Development and Global Citizenship"; "Education for Sustainable Development in the UK: Making the Connections between the Environment and Development Agendas," *Theory and Research in Education* 6, no. 2 (2008); and Philip Collie, *Barriers and Motivators for Adopting Sustainability Programmes in Schools* (Cheltenham, UK: Schoolzone, 2008).

17. Noah Feinstein, *Education for Sustainable Development in the United States of America: A Report Submitted to the International Alliance of Leading Education Institutes* (Madison: University of Wisconsin, 2009); Charles Hopkins, "Education for Sustainable Development in Formal Education in Canada," in *Schooling for Sustainable Development in Canada and the United States*, ed. Rosalyn McKeown and Victor Nolet (New York: Springer, 2013); and Kim Smith et al., *The Status of Education for*

Sustainable Development (ESD) in the United States: A 2015 Report to the US Department of State, International Society of Sustainability Professionals, December 2015, https://www.sustainabilityprofessionals.org/sites/default/files/ESD%20in%20the%20United%20States%20final.pdf.

18. Peggy Bartlett and Geoffrey Chase, eds., *Sustainability on Campus: Stories and Strategies for Change* (Cambridge, MA: MIT Press, 2004); Everett, "Sustainability in Higher Education"; David Orr, "What Is Higher Education for Now?," in *State of the World 2010: Transforming Cultures, from Consumerism to Sustainability*, ed. Worldwatch Institute (New York: W. W. Norton, 2010); and Paula Jones, David Selby, and Stephen Sterling, *Sustainability Education: Perspectives and Practice across Higher Education* (London: Earthscan, 2010). Cf. Cheryl Desha and Karlson Hargroves, *Engineering Education and Sustainable Development: A Guide to Rapid Curriculum Renewal in Higher Education* (London: Earthscan, 2011); and Smith et al., *Status of Education for Sustainable Development in the United States*.

19. UNESCO, *Shaping the Future We Want: UN Decade of Education for Sustainable Development (2005–2014) Final Report* (Paris: UNESCO, 2014), 9, http://unesdoc.unesco.org/images/0023/002301/230171e.pdf.

20. UNESCO, *Roadmap for Implementing the Global Action Programme on Education for Sustainable Development* (Paris: UNESCO, 2014), http://unesdoc.unesco.org/images/0023/002305/230514e.pdf.

21. Kim Smith et al., "UNESCO Roadmap for Implementing the Global Action Programme on Education for Sustainable Development: Implementation Recommendations for the United States of America" (US Delegation to the UNESCO World Conference on ESD, 2015), http://gpsen.org/wp-content/uploads/2016/05/GAP-Roadmap-Recommendations-Final.pdf.

22. Kim Smith et al., *The Status of Education for Sustainable Development (ESD) in the United States*.

23. Debra Rowe, Susan Jane Gentile, and Lilah Clevey, "The US Partnership for Education for Sustainable Development: Progress and Challenges Ahead," *Applied Environmental Education & Communication* 14, no. 2 (2015).

24. Carmela Federico and Jamie Cloud, "Kindergarten through Twelfth Grade Education: Fragmentary Progress in Equipping Students to Think and Act in a Challenging World," in *Agenda for a Sustainable America*, ed. John Dernbach (Washington, DC: Environmental Law Institute Press, 2009); and Noah Feinstein and Ginny Carlton, "Education for Sustainability in the K–12 Educational System of the United States," in *Schooling for Sustainable Development in Canada and the United States*, ed. Rosalyn McKeown and Victor Nolet (New York: Springer, 2013).

25. Victor Nolet, "Preparing Sustainability-Literate Teachers," *Teachers College Record* 111, no. 2 (2009).

26. Pub. L. 107–110, 115 STAT 1425 (January 8, 2002).

27. White House Press Office, "White House Report: The Every Child Succeeds Act," The White House, Office of the Press Secretary, December 10, 2015, https://www. whitehouse.gov/the-press-office/2015/12/10/white-house-report-every-student-succeeds-act; Sarah Bodor, "Every Student Succeeds Act Includes Historic Gains for Environmental Education," NAAEE, December 9, 2015, https://naaee.org/eepro/resources/every-student-succeeds-act-includes-historic-gains-environmental-education; and National Research Council, *Education for Life and Work: Developing Transferable Knowledge and Skills in the 21st Century* (Washington, DC: National Academies Press, 2012).

28. NRC, *Education for Life and Work.*

29. Ibid.

30. Ibid.

31. Ibid.

32. Ibid.

33. See NGSS Lead States, Next Generation Science Standards website (NGSS Lead States, 2013), http://www.nextgenscience.org/.

34. National Research Council, *A Framework for K–12 Science Education: Practices, Crosscutting Concepts, and Core Ideas* (Washington, DC: The National Academies Press, 2012), http://www.nd.edu/~nismec/articles/framework-science%20standards .pdf.

35. See NGSS Lead States, Next Generation Science Standards website; and Michael E. Wysession, "The 'Next Generation Science Standards' and the Earth and Space Sciences," *Science and Children* 50, no. 8 (April 2013), http://eric.ed.gov/?id=EJ1020542.

36. See NGSS Lead States, Next Generation Science Standards website; and Wysession, "'Next Generation Science Standards.'"

37. National Research Council, *A Framework for K–12 Science Education.*

38. Michael E. Wysession, "Implications for Earth and Space in New K–12 Science Standards," *Eos, Transactions, American Geophysical Union* 93, no. 46 (2012); and Wysession, "'Next Generation Science Standards.'"

39. Noah Feinstein and Kathryn L. Kirchgasler, "Sustainability in Science Education? How the Next Generation Standards Approach Sustainability, and Why It Matters," *Science Education* 99, no. 1 (2015): 121.

40. Sarah Galey, "Education Politics and Policy: Emerging Institutions, Interests, and Ideas," *Policy Studies Journal: The Journal of the Policy Studies Organization* 43, no. 1 (2015).

41. Valerie Strauss, "Revolt against High-Stakes Standardized Testing Growing—and So Does Its Impact," *Washington Post*, March 19, 2015, http://www.washingtonpost.com/blogs/answer-sheet/wp/2015/03/19/revolt-against-high-stakes-standardized-testing-growing-and-so-does-its-impact/.

42. Elaine McArdle, "What Happened to the Common Core?," *Harvard Ed. Magazine*, September 3, 2014, http://www.gse.harvard.edu/news/ed/14/09/what-happened-common-core.

43. Nicole D. LaDue, "Help to Fight the Battle for Earth in US Schools," *Nature* 519, no. 7542 (2015).

44. UNESCO, *Shaping the Future We Want*.

45. The significance of sustainability for development is acknowledged in Potsdam Institute for Climate Impact Research and Climate Analysis, *Turn Down the Heat*.

46. Bourn, "Education for Sustainable Development and Global Citizenship"; "Education for Sustainable Development in the UK."

47. W. Scott, "Education and Sustainable Development: Challenges, Responsibilities, and Frames of Mind," *Trumpeter* 18, no. 1 (2002); Bob Jickling and Arjen E. J. Wals, "Globalization and Environmental Education: Looking beyond Sustainable Development," *Journal of Curriculum Studies* 40, no. 1 (2007); and David Selby, "The Firm and Shaky Ground of Education for Sustainable Development."

48. Oxfam's Education for Global Citizenship (EGC) curriculum framework places a heavy emphasis on readiness to advance global justice. Early versions of it did not reflect much concern with the curricular basis for understanding problems of sustainability, but the latest version includes an encouraging cross-curricular approach to sustainable development, using Philosophy for Children to teach critical thinking (Oxfam, *Education for Global Citizenship: A Guide for Schools* [Oxfam, 2015], http://www.oxfam.org.uk/education/global-citizenship/global-citizenship-guides). The Oxfam EGC curriculum has a significant presence in schools in the United Kingdom (Bourn, "Education for Sustainable Development in the UK," 11–12).

49. Bob Jickling, "Why I Don't Want My Children to Be Educated for Sustainable Development," *Journal of Environmental Education* 23, no. 4 (1992); Scott, "Education and Sustainable Development"; Christopher Schlottmann, "Educational Ethics and the DESD: Considering the Trade-Offs," *Theory and Research in Education* 6, no. 2 (2008); and Spearman, "Sustainability Education."

50. Andrew Dobson, "Environmental Sustainabilities: An Analysis and a Typology," *Environmental Politics* 5, no. 3 (1996); Sharachchandra Lélé, "Sustainable Development: A Critical Review," in *Environment: An Interdisciplinary Anthology*, ed. Glenn Adelson et al. (New Haven, CT: Yale University Press, 2008); and Bonevac, "Is Sustainability Sustainable?"

51. Jickling and Wals, "Globalization and Environmental Education."

52. Bourn, "Education for Sustainable Development in the UK"; and Schlottmann, "Educational Ethics and the DESD."

53. Jickling, "Why I Don't Want My Children Educated for Sustainable Development"; J. Smyth, *Are Educators Ready for the Next Earth Summit?* (London: Stakeholder Forum, 2002); and Scott, "Education and Sustainable Development."

54. Jickling and Wals, "Globalization and Environmental Education," 5.

55. Scott, "Education and Sustainable Development"; Jickling and Wals, "Globalization and Environmental Education."

56. Schlottmann, "Educational Ethics and the DESD."

57. UNESCO, Education for Sustainable Development: United Nations Decade (2005–2014) website (UNESCO, 2005), http://en.unesco.org/themes/education -sustainable-development.

58. Ibid.

59. See chapter 2.

60. Randall Curren, "Cultivating the Moral and Intellectual Virtues," in *Philosophy of Education: An Anthology* (Oxford: Blackwell Publishing, 2007); "Judgment and the Aims of Education"; "Motivational Aspects of Moral Learning and Progress"; "Virtue Ethics and Moral Education."

61. UNESCO, Education for Sustainable Development: United Nations Decade (2005–2014) website (UNESCO, 2005), http://en.unesco.org/themes/education -sustainable-development, under Quality Education.

62. Regarding the ESD Toolkit 2.0, see Rosalyn McKeown, *Education for Sustainable Development Toolkit, Version 2.0* (2007), http://www.esdtoolkit.org/discussion/ reorient.htm, under Reorienting Education. On the need to rethink orthodox development economics, see Sen, *Development as Freedom*; and Stiglitz, Sen, and Fitousi, *Mismeasuring Our Lives*.

63. Gerald Nosich, *Learning to Think Things Through: A Guide to Critical Thinking across the Curriculum*, 4th ed. (Upper Saddle River, NJ: Pearson, 2011), provides a model for how to do this. It provides students with tools to identify the basic structure of any fields they might study and bring the resources of those fields together in practical reasoning and judgment.

64. See Harvey Siegel, *Educating Reason* (New York: Routledge, 1988); and Matthew Lipman, *Thinking in Education*, 2nd ed. (Cambridge: Cambridge University Press, 2003).

65. For an account of the global reach of economic relations, see Mike Davis, *Planet of Slums* (London: Verso, 2009). On the global reach of pollution and its impact on health and food sources, see Dodds, *Humanity's Footprint*, 3, 48–62; and Guidotti, *Health and Sustainability*.

66. W. Neil Adger et al., eds., *Fairness in Adaptation to Climate Change* (Cambridge, MA: MIT Press, 2006); and Moser and Boykoff, eds., *Successful Adaptation to Climate Change*.

67. As noted in the opening section of this chapter, environmental "literacy" is understood broadly to include "the capacity to identify root causes of threats to sustainable development and the values, motivations and skills to address them." This would surely include a capacity to recognize when wars are being contemplated and fought out of unwillingness to accept just and peaceful distributions of scarce resources.

68. Schor, *Plenitude*.

69. Diamond, *Collapse*, 211–276.

70. Joy Horowitz, *The Poisoning of an American High School* (New York: Penguin, 2007), depicts a stunning missed learning opportunity of this kind.

71. See Curren and Dorn, *Patriotic Education*.

72. See Randall Curren, "Moral Education and Juvenile Crime," in *Nomos XLIII: Moral and Political Education*, ed. Stephen Macedo and Yael Tamir (New York: NYU Press, 2002); and "A Neo-Aristotelian Account of Education, Justice, and the Human Good," *Theory and Research in Education* 11, no. 3 (2013).

73. See Robert Fullinwider, "Moral Conventions and Moral Lessons," *Social Theory and Practice* 15, no. 3 (1989); Michael Pritchard, *Reasonable Children: Moral Education and Moral Learning* (Lawrence: University Press of Kansas, 1996).

74. A model that has gained popularity with students in recent years is the National High School Ethics Bowl (http://nhseb.unc.edu/).

75. See Curren, "Motivational Aspects of Moral Learning and Progress."

76. John Kleinig identifies authority as a form of influence resting on the perception that someone knows what is to be done (in some sense); Kleinig, *Philosophical Issues in Education* (London: Croom Helm, 1982), 213.

77. See Joan Goodman, "Student Authority: Antidote to Alienation," *Theory and Research in Education* 8, no. 3 (2010).

78. Ibid., 241 (italics added).

Conclusion

1. Hannah Arendt, *Between Past and Future* (London: Faber & Faber, 1961), 185–186. Arendt wrote these words not with the ravaging of Earth in mind, but the ravaging of Europe by totalitarian ideologies. In the context of current ecological knowledge, her ideas of nature and earth alienation might have lent her words a wider meaning.

Bibliography

Adelson, Glenn, James Engell, Brent Ranalli, and K. P. Van Anglen, eds. *Environment: An Interdisciplinary Anthology*. New Haven, CT: Yale University Press, 2008.

Adger, W. Neil, Jouni Paavola, Saleemul Huq, and M. J. Mace, eds. *Fairness in Adaptation to Climate Change*. Cambridge, MA: MIT Press, 2006.

AghaKouchak, Amir, David Feldman, Michael J. Stewardson, Jean-Daniel Saphores, Stanley Grant, and Brett Sanders. "Australia's Drought: Lessons for California." *Science* 343, no. 6178 (2014): 1430–1431.

Agrawal, Arun. "Common Resources and Institutional Sustainability." In *The Drama of the Commons*, edited by Elinor Ostrom, Thomas Dietz, Nives Solsak, Paul C. Stern, Susan Stonich, and Elke U. Weber, 41–86. Washington, DC: National Academies Press, 2002.

Ainslie, George. *Breakdown of Will*. Cambridge: Cambridge University Press, 2001.

Allen, T. F. H., Joseph Tainter, and Thomas W. Hoekstra. *Supply-Side Sustainability*. New York: Columbia University Press, 2003.

Allouche, Jeremy, Carl Middleton, and Dipak Gyawali. "Nexus Nirvana or Nexus Nullity? A Dynamic Approach to Security and Sustainability in the Water-Energy-Food Nexus." STEPS Working Paper 63, STEPSCentre, Brighton, UK, 2014. http://steps-centre.org/wp-content/uploads/Water-and-the-Nexus.pdf.

Andreou, Chrisoula. "Environmental Preservation and Second-Order Procrastination." *Philosophy & Public Affairs* 35, no. 3 (2007): 233–248.

Andreou, Chrisoula. "Understanding Procrastination." *Journal for the Theory of Social Behaviour* 37 (2007): 183–193.

Andrews, Cecile. *The Circle of Simplicity*. New York: Harper, 1997.

Annas, Julia. *Intelligent Virtue*. Oxford: Oxford University Press, 2011.

Archer, David. *The Long Thaw: How Humans Are Changing the Next 100,000 Years of Earth's Climate*. Princeton, NJ: Princeton University Press, 2009.

Arendt, Hannah. *Between Past and Future*. London: Faber & Faber, 1961.

Armitage, Derek, Rob C. de Loë, Michelle Morris, Tom W.D. Edwards, Andrea K. Gerlak, Roland I. Hall, Dave Huitema, Ray Ison, David Livingstone, Glen MacDonald, Naha Mirumachi, Ryan Plummer, and Brent B. Wolfe. "Science-Policy Processes for Transboundary Water Governance." *AMBIO* 44, no. 5 (2015): 353–366.

Aşıcı, Ahmet Atıl, and Sevil Acar. "Does Income Growth Relocate Ecological Footprint?" *Ecological Indicators* 61 (2016): 707–714.

Associated Press. "5 Years after BP Spill, Drillers Push into Riskier Depths." *Chicago Tribune*, April 20, 2015. http://www.chicagotribune.com/news/nationworld/chi-bp -oil-spill-20150419-story.html.

Attas, Daniel. "A Transgenerational Difference Principle." In *Intergenerational Justice*, edited by Axel Gosseries and Lukas H. Meyer, 189–218. Oxford: Oxford University Press, 2009.

Australian Public Service Commission. *Tackling Wicked Problems: A Public Policy Perspective*. Canberra: Australian Public Service Commission, 2007.

Bäckstrand, Karin. "The Democratic Legitimacy of Global Governance after Copenhagen." In *The Oxford Handbook of Climate Change and Society*, edited by John S. Dryzek, Richard B. Norgaard, and David Schlosberg, 669–684. Oxford: Oxford University Press, 2011.

Baer, Paul, Tom Athanasiou, Simon Kartha, and Eric Kemp-Benedict. "Greenhouse Development Rights: A Framework for Climate Protection That Is 'More Fair' than Equal Per Capita Emissions Rights." In *Climate Ethics*, edited by Stephen M. Gardiner, Simon Caney, Dale Jamieson, and Henry Shue, 213–230. Oxford: Oxford University Press, 2010.

Bager, Simon. "Big Facts: Focus on East and Southeast Asia." CGIAR, May 6, 2014. http://ccafs.cgiar.org/blog/big-facts-focus-east-and-southeast-asia#.VQc6ZY7F98E.

Baker, David P. "The Educational Transformation of Work: Towards a New Synthesis." *Journal of Education and Work* 22 (2009): 163–193.

Baker, David P. "Forward and Backward, Horizontal and Vertical: Transformation of Occupational Credentialing in the Schooled Society." *Research in Social Stratification and Mobility* 29 (2011): 5–29.

Baker, David P. *The Schooled Society: The Educational Transformation of Global Culture*. Stanford, CA: Stanford University Press, 2014.

Balot, Ryan. *Greed and Injustice in Classical Athens*. Princeton, NJ: Princeton University Press, 2001.

Barlow, Maude. *Blue Future: Protecting Water for People and the Planet Forever*. New York: The New Press, 2013.

Barnosky, Anthony D., James H. Brown, Gretchen C. Daily, Rodolfo Dirzo, Anne H. Ehrlich, Paul R. Ehrlich, and Jussi T. Eronen. "Introducing the Scientific Consensus on Maintaining Humanity's Life Support Systems in the 21st Century: Information for Policy Makers." *The Anthropocene Review* 1, no. 1 (2014): 78–109.

Barry, Brian. "Sustainability and Intergenerational Justice." In *Environmental Ethics*, edited by Andrew Light and Holmes Rolston III, 487–499. Malden, MA: Blackwell Publishing, 2003.

Barry, Brian. *Why Social Justice Matters*. Cambridge: Polity Press, 2005.

Bartlett, Peggy, and Geoffrey Chase, eds. *Sustainability on Campus: Stories and Strategies for Change*. Cambridge, MA: MIT Press, 2004.

Batie, Sandra S. "Wicked Problems and Applied Economics." *American Journal of Agricultural Economics* 90, no. 5 (2008): 1176–1191.

Bea, Robert. "Understanding the Macondo Well Failures." Deepwater Horizon Study Group Working Paper, January 2011. http://ccrm.berkeley.edu/pdfs_papers/DH SGWorkingPapersFeb16-2011/UnderstandingMacondoWellFailures-BB_DHSG -Jan2011.pdf.

Becker, Christian. *Sustainability Ethics and Sustainability Research*. Dordrecht, Netherlands: Springer, 2011.

Bellamy, Richard. *Citizenship: A Very Short Introduction*. Oxford: Oxford University Press, 2008.

Biello, David. *The Unnatural World: The Race to Remake Civilization in Earth's Newest Age*. New York: Scribner, 2016.

Bills, David. *Sociology of Education and Work*. Malden, MA: Blackwell, 2004.

Bills, David, and David Brown, eds. *New Directions in Educational Credentialism*. Special issue, *Research in Social Stratification and Mobility* 29, no. 1 (2011): 1–138.

Blomqvist, L., B. W. Brook, E. C. Ellis, P. M. Kareiva, T. Nordhaus, and M. Shellenberger. "Does the Shoe Fit? Real versus Imagined Ecological Footprints." *PLoS Biology* 11, no. 11 (2013). doi:10.1371/journal.pbio.1001700.

Bodor, Sarah. "Every Student Succeeds Act Includes Historic Gains for Environmental Education." NAAEE, December 9, 2015. https://naaee.org/eepro/resources/every -student-succeeds-act-includes-historic-gains-environmental-education.

Bonevac, Daniel. "Is Sustainability Sustainable?" *Academic Questions* 23 (2010): 84–101.

Boudon, Raymond. *Education, Opportunity, and Social Inequality: Changing Prospects in Western Society*. New York: Wiley, 1974.

Bourdieu, Pierre, and Jean-Claude Passeron. *Reproduction in Education, Society, and Culture.* Beverly Hills, CA: Sage Publications, 1977.

Bourn, Douglas. "Education for Sustainable Development and Global Citizenship: The UK Perspective." *Applied Environmental Education and Communication* 4, no. 3 (2005): 233–237.

Bourn, Douglas. "Education for Sustainable Development in the UK: Making the Connections between the Environment and Development Agendas." *Theory and Research in Education* 6, no. 2 (2008): 193–206.

Bowen, Frances. *After Greenwashing: Symbolic Corporate Environmentalism and Society.* Cambridge: Cambridge University Press, 2014.

Bowles, Samuel, and Herbert Gintis. *Schooling in Capitalist America.* New York: Basic Books, 1976.

"BP's Troubled Past." *PBS Frontline*, October 26, 2010. http://www.pbs.org/wgbh/pages/frontline/the-spill/bp-troubled-past/.

Breyfogle, Nicholas. "Dry Days Down Under: Australia and the World Water Crisis." *Origins* 3, no. 7 (2010). http://origins.osu.edu/article/dry-days-down-under-australia-and-world-water-crisis.

Brighouse, Harry. "Globalization and the Professional Ethic of the Professoriat." In *Global Inequalities and Higher Education*, edited by Elaine Unterhalter and Vincent Carpentier, 287–311. New York: Palgrave Macmillan, 2010.

Brighouse, Harry, and Ingrid Robeyns, eds. *Measuring Justice: Primary Goods and Capabilities.* Cambridge: Cambridge University Press, 2010.

Brock, Gillian. *Global Justice: A Cosmopolitan Approach.* Oxford: Oxford University Press, 2009.

Broome, John. *Climate Matters: Ethics in a Warming World.* New York: Norton, 2012.

Brown, David K. *Degrees of Control: A Sociology of Educational Expansion and Occupational Credentialism.* New York: Teachers College Press, 1995.

Brown, Elaine M. "The Deepwater Horizon Disaster." In *Case Studies in Organizational Communication: Ethical Perspectives and Practices*, ed. Steve May, 233–246. Thousand Oaks, CA: Sage Press, 2012.

Brown, Marilyn, Jess Chandler, Melissa V. Lapsa, and Benjamin K. Sovacool. *Carbon Lock-In: Barriers to Deploying Climate Change Mitigation Technologies.* Oak Ridge, TN: Oak Ridge National Laboratory, 2007.

Brown, Peter G., and Jeremy J. Schmidt, eds. *Water Ethics: Foundational Readings for Students and Professionals.* Washington, DC: Island Press, 2010.

Buchanan, Allen. "Political Liberalism and Social Epistemology." *Philosophy & Public Affairs* 32, no. 2 (2004): 95–130.

Buchanan, Allen. "Social Moral Epistemology." *Social Philosophy and Policy* 19 (2002): 126–152.

Bush, George W. "State of the Union Address," Washington, DC, 2007.

Campbell, Ian, Barry Hart, and Chris Barlow. "Integrated Management in Large River Basins: 12 Lessons from the Mekong and Murray-Darling Rivers." *River Systems* 20, no. 3–4 (2013): 231–247.

Caradonna, Jeremy L. *Sustainability: A History.* Oxford: Oxford University Press, 2014.

Cardenas, Juan-Camilo, Luz Angela Rodriguez, and Nancy Johnson. "Vertical Collective Action: Addressing Vertical Asymmetries in Watershed Management." CEDE, February 2015. doi:10.13140/RG.2.1.2701.7767. https://www.researchgate .net/publication/275351883_Vertical_Collective_Action_Addressing_Vertical _Asymmetries_in_Watershed_Management.

Castree, Noel. "Reply to 'Strategies for Changing the Intellectual Climate' and 'Power in Climate Change Research.'" *Nature Climate Change* 5 (May 2015): 393.

Castree, Noel, William M. Adams, John Barry, Daniel Brockington, Bram Büscher, Esteve Corbera, David Demeritt, Rosaleen Duffy, Ulrike Felt, and Katja Neves. "Changing the Intellectual Climate." *Nature Climate Change* 4 (September 2014): 763–768.

Cavanaugh, John, and Jerry Mander, eds. *Alternatives to Global Capitalism: A Better World Is Possible.* San Francisco: Berrett-Koehler, 2002.

Chambers, Nicky, Craig Simmons, and Mathias Wackernagel. *Sharing Nature's Interest: Ecological Footprints as an Indicator of Sustainability.* London: Earthscan, 2000.

Chirkov, Valery I., Richard M. Ryan, and Kennon M. Sheldon, eds. *Human Autonomy in Cross-Cultural Context: Perspectives on the Psychology of Agency, Freedom, and Well-Being.* Dordrecht, Netherlands: Springer, 2011.

Church, Wendy, and Laura Skelton. "Infusing Sustainability across the Curriculum." In *Schooling for Sustainable Development in Canada and the United States,* edited by Rosalyn McKeown and Victor Nolet, 183–195. New York: Springer, 2013.

Cleveland, Cutler J., C. Michael Hogan, and Peter Saundry. "Deepwater Horizon Oil Spill." In *Encyclopedia of Earth,* edited by Cutler J. Cleveland. Washington, DC: Environmental Information Coalition, National Council for Science and the Environment, 2010. https://eoearthlive.wordpress.com/.

Cohen, Stanley. *States of Denial.* Cambridge: Polity Press, 2001.

Collie, Philip. *Barriers and Motivators for Adopting Sustainability Programmes in Schools.* Cheltenham, UK: Schoolzone, 2008.

Collins, Randall. *The Credential Society: An Historical Sociology of Education and Stratification.* New York: Academic Press, 1979.

Columbia Accident Investigation Board. *Report of the Columbia Accident Investigation Board.* Washington, DC: National Aeronautics and Space Administration, 2003.

Committee on the Affordability of National Insurance Program Premiums. *Affordability of National Flood Insurance Program Premiums. Report 1.* Washington, DC: National Academies Press, 2015. http://www.nap.edu/catalog/21709/affordability-of-national-flood-insurance-program-premiums-report-1.

Committee on the Human Dimensions of Global Change, Elinor Ostrom, Thomas Dietz, Nives Dolšak, Paul C. Stern, Susan Stonich, and Elke U. Weber, eds. *The Drama of the Commons.* Washington, DC: National Academies Press, 2002.

Conca, Ken. "The Rise of the Region in Global Environmental Governance." *Global Environmental Politics* 12, no. 3 (August 2012): 127–133.

Condon, Patrick. *Seven Rules for Sustainable Communities: Design Strategies for the Post-Carbon World.* Washington, DC: Island Press, 2010.

Conklin, Jeff. *Wicked Problems and Social Complexity.* Napa, CA: CogNexus Institute, 2006.

Connor, Steve. "The State of the World? It is on the Brink of Disaster." *Independent,* March 30, 2005. http://www.independent.co.uk/news/science/the-state-of-the-world-it-is-on-the-brink-of-disaster-530432.html.

Cooley, Heather, Newsha Ajami, Mai-Lan Ha, Veena Srinivasan, Jason Morrison, Kristina Donnelly, and Juliet Christian-Smith. *Global Water Governance in the 21st Century.* Oakland, CA: Pacific Institute, 2013.

Cooper, John, ed. *Plato: Complete Works.* Indianapolis: Hackett, 1997.

Cooper, Mark. "The Economic and Institutional Foundations of the Paris Agreement on Climate Change: The Political Economy of Roadmaps to a Sustainable Electricity Future." January 26, 2016. http://papers.ssrn.com/sol3/Papers.cfm?abstract_id=2722880.

Cornwall, Warren. "Deepwater Horizon: After the Oil." *Science* 348, no. 6230 (2015): 22–29.

Costanza, Robert, and Herman Daly. "Natural Capital and Sustainable Development." *Conservation Biology* 6, no. 1 (1992): 37–46.

Costanza, Robert, Lisa J. Graumlich, and Will Steffen, eds. *Sustainability or Collapse? An Integrated History and Future of People on Earth.* Cambridge, MA: MIT Press, 2011.

Crocker, David, and Toby Linden, eds. *Ethics of Consumption: The Good Life, Justice, and Global Stewardship.* Lanham, MD: Rowman & Littlefield, 1998.

Curren, Randall. "Aristotelian Necessities." *The Good Society* 22, no. 2 (Fall 2013): 247–263.

Curren, Randall. *Aristotle on the Necessity of Public Education.* Lanham, MD: Rowman & Littlefield, 2000.

Curren, Randall, ed. *A Companion to the Philosophy of Education.* Oxford: Blackwell, 2003.

Curren, Randall. "Cultivating the Moral and Intellectual Virtues." In *Philosophy of Education: An Anthology,* edited by Randall Curren, 507–516. Oxford: Blackwell Publishing, 2007.

Curren, Randall. "Defining Sustainability Ethics." In *Environmental Ethics,* 2nd ed., edited by Michael Boylan, 331–345. Oxford: Wiley-Blackwell, 2013.

Curren, Randall. *Education for Sustainable Development: A Philosophical Assessment.* London: PESGB, 2009.

Curren, Randall. "Judgment and the Aims of Education." *Social Philosophy & Policy* 31, no. 1 (Fall 2014): 36–59.

Curren, Randall. "Meaning, Motivation, and the Good." Professorial Inaugural Lecture, Royal Institute of Philosophy, London, January 24, 2014. http://www .youtube.com/watch?v=rhjZvbvpJYQ&feature=youtu.be.

Curren, Randall. "Moral Education and Juvenile Crime." In *Nomos XLIII: Moral and Political Education,* edited by Stephen Macedo and Yael Tamir, 359–380. New York: NYU Press, 2002.

Curren, Randall. "Motivational Aspects of Moral Learning and Progress." *Journal of Moral Education* 43, no. 4 (December 2014): 484–499.

Curren, Randall. "A Neo-Aristotelian Account of Education, Justice, and the Human Good." *Theory and Research in Education* 11, no. 3 (2013): 232–250.

Curren, Randall, ed. *Philosophy of Education: An Anthology.* Oxford: Blackwell Publishing, 2007.

Curren, Randall. "Sustainability in the Education of Professionals." *Journal of Applied Ethics and Philosophy* 2 (September 2010): 21–29.

Curren, Randall. "Virtue Ethics and Moral Education." In *Routledge Companion to Virtue Ethics,* edited by Michael Slote and Lorraine Besser-Jones, 459–470. London: Routledge, 2015.

Curren, Randall. "A Virtue Theory of Moral Motivation." Paper presented at the Varieties of Virtue Ethics in Philosophy, Social Science and Theology Conference, Oriel College, Oxford, January 8–10, 2015. http://www.jubileecentre.ac.uk/userfiles/jubileecentre/pdf/conference-papers/Varieties_of_Virtue_Ethics/Curren_Randall.pdf.

Curren, Randall, and Chuck Dorn. *Patriotic Education in a Global Age*. Chicago: University of Chicago Press, 2017.

Daily, Gretchen C., Stephen Polasky, Joshua Goldstein, Peter M. Kareiva, Harold A. Mooney, Liba Pejar, Taylor H. Ricketts, James Salzman, and Robert Shellenberger. "Ecosystem Services in Decision Making: Time to Deliver." *Frontiers in Ecology and the Environment* 7, no. 1 (2009): 21–28.

Davenport, Coral. "U.S. Will Allow Drilling for Oil in Arctic Ocean." *New York Times*, May 11, 2015. http://www.nytimes.com/2015/05/12/us/white-house-gives -conditional-approval-for-shell-to-drill-in-arctic.html.

Davis, Mike. *Planet of Slums*. London: Verso, 2009.

De Wall, Franz. *The Age of Empathy: Nature's Lessons for a Kinder Society*. New York: Random House, 2009.

Deci, Edward L., Haleh Eghrani, Brian C. Patrick, and Dean R. Leone. "Facilitating Internalization: The Self-Determination Theory Perspective." *Journal of Personality* 62, no. 1 (1994): 119–142.

Deci, Edward L., Jennifer G. La Guardia, Arlen C. Moller, Marc J. Scheiner, and Richard M. Ryan. "On the Benefits of Giving as well as Receiving Autonomy Support: Mutuality in Close Friendships." *Personality and Social Psychology Bulletin* 32, no. 3 (2006): 313–327.

Deci, Edward L., and Richard M. Ryan. "Motivation, Personality, and Development within Embedded Social Contexts: An Overview of Self-Determination Theory." In *The Oxford Handbook of Human Motivation*, edited by Richard Ryan, 85–107. New York: Oxford University Press, 2012.

Deepwater Horizon Study Group. *Final Report on the Investigation of the Macondo Well Blowout*. March 1, 2011. http://www.aspresolver.com/aspresolver.asp?ENGV ;2082362.

Department for Children, Schools and Families. *Brighter Futures—Greener Lives: Sustainable Development Action Plan 2008–2010*. 2008. http://www.unece.org/fileadmin/ DAM/env/esd/Implementation/NAP/UK.SDActionPlan.e.pdf.

Department of the Environment, Water, Heritage and the Arts. *Living Sustainably: The Australian Government's National Action Plan for Education for Sustainability*. Canberra: Department of the Environment, Water, Heritage and the Arts, 2009.

Dernbach, John, ed. *Agenda for a Sustainable America*. Washington, DC: Environmental Law Institute Press, 2009.

Desha, Cheryl, and Karlson Hargroves. *Engineering Education and Sustainable Development: A Guide to Rapid Curriculum Renewal in Higher Education*. London: Earthscan, 2011.

de Vries, Bert J. M. *Sustainability Science*. Cambridge: Cambridge University Press, 2013.

Diamond, Jared. *Collapse: How Societies Choose to Fail or Succeed*. New York: Viking, 2005.

Diamond, Jared. "Invention Is the Mother of Necessity." *New York Times Magazine*, 1999. http://partners.nytimes.com/library/magazine/millennium/m1/diamond.html.

Dietz, Thomas. "Elinor Ostrom: 1933–2012." *Solutions* 3, no. 5 (August 2012): 6–7. http://www.thesolutionsjournal.com/node/1166.

Dietz, Thomas. "Informing Sustainability Science through Advances in Environmental Decision Making and Other Areas of Science." Paper presented at the NRC Sustainability Science Roundtable, Irvine, CA, January 14–15, 2016. http://sites.nationalacademies.org/cs/groups/pgasite/documents/webpage/pga_170344.pdf.

Dietz, Thomas. "Prolegomenon to a Structural Human Ecology of Human Well-Being." *Sociology of Development* 1 (2015): 123–148.

Dietz, Thomas, Gerald T. Gardner, Jonathan Gilligan, Paul C. Stern, and Michael P. Vandenbergh. "Household Actions Can Provide a Behavioral Wedge to Rapidly Reduce US Carbon Emissions." *Proceedings of the National Academy of Sciences of the United States of America* 106, no. 44 (November 2009): 18452–18456.

Dietz, Thomas, Eugene A. Rosa, and Richard York. "Driving the Human Ecological Footprint." *Frontiers in Ecology and the Environment* 5 (2007): 13–18.

Dietz, Thomas, Eugene A. Rosa, and Richard York. "Environmentally Efficient Well-Being: Is There a Kuznets Curve?" *Applied Geography* 32, no. 1 (2012): 21–28.

Dietz, Thomas, Eugene A. Rosa, and Richard York. "Environmentally Efficient Well-Being: Rethinking Sustainability as the Relationship between Human Well-Being and Environmental Impacts." *Human Ecology Review* 16 (2009): 113–122.

Dobson, Andrew. *Citizenship and the Environment*. Oxford: Oxford University Press, 2003.

Dobson, Andrew. "Environmental Sustainabilities: An Analysis and a Typology." *Environmental Politics* 5, no. 3 (1996): 401–428.

Dobson, Andrew, and Derek Bell, eds. *Environmental Citizenship*. Cambridge, MA: MIT Press, 2006.

Dodds, Walter. *Humanity's Footprint: Momentum, Impact, and Our Global Environment.* New York: Columbia University Press, 2008.

Dore, Ronald. *The Diploma Disease: Education, Qualification, and Development.* London: Allen and Unwin, 1976.

Dryzek, John. *The Politics of the Earth: Environmental Discourses.* 3rd ed. New York: Oxford University Press, 2013.

Dryzek, John. *Rational Ecology: Environment and Political Economy.* Oxford: Blackwell, 1987.

Dryzek, John S., Richard B. Norgaard, and David Schlosberg, eds. *The Oxford Handbook of Climate Change and Society.* Oxford: Oxford University Press, 2011.

Dunlap, Riley E., and Aaron M. McCright. "Organized Climate Change Denial." In *The Oxford Handbook of Climate Change and Society,* edited by John S. Dryzek, Richard B. Norgaard, and David Schlosberg, 144–160. Oxford: Oxford University Press, 2011.

Eastman, A. D. "The Homeowner Flood Insurance Affordability Act: Why the Federal Government Should Not Be in the Insurance Business." *American Journal of Business and Management* 4, no. 2 (2015): 71–75.

Echegaray, Jacqueline Nolley, and Carol Elizabeth Lockwood. "Reproductive Rights Are Human Rights." In *A Pivotal Moment,* edited by Laurie Mazur, 341–352. Washington, DC: Island Press, 2010.

Edwards, Andres. *The Sustainability Revolution.* Gabriola Island, British Columbia: New Society Publishers, 2005.

Egan, Timothy. *The Worst Hard Time.* New York: Mariner Books, 2006.

Eilperin, Juliet. "Carbon Output Must Near Zero to Avert Danger, New Studies Say." *Washington Post,* March 10, 2008, A01.

Elgin, Duane. *Voluntary Simplicity.* 2nd rev. ed. New York: Harper, 2010.

Elliott, Debbie. "Five Years after BP Oil Spill, Effects Linger and Recovery Is Slow." *NPR,* April 20, 2015. http://www.npr.org/2015/04/20/400374744/5-years-after-bp-oil-spill-effects-linger-and-recovery-is-slow.

Elster, Jon. *Ulysses Unbound: Studies in Rationality, Precommitment, and Constraints.* Cambridge: Cambridge University Press, 2000.

Engineering Council UK. *Guidance on Sustainability for the Engineering Profession.* London: Engineering Council UK, 2009. http://www.engc.org.uk/engcdocuments/internet/Website/Guidance%20on%20Sustainability.pdf.

Engle, Nathan L., Owen R. Johns, Maria Carmen Lemos, and Donald R. Nelson. "Integrated and Adaptive Management of Water Resources: Tensions, Legacies, and the Next Best Thing." *Ecology and Society* 16, no. 1 (2011): 19. http://www .ecologyandsociety.org/vol16/iss1/art19/.

Erlanger, Steven. "With Prospect of U.S. Slowdown, Europe Fears a Worsening Debt Crisis." *New York Times*, August 8, 2011, B3.

Everett, Jennifer. "Sustainability in Higher Education: Implications for the Disciplines." *Theory and Research in Education* 6, no. 2 (2008): 237–251.

FAO. *World Review of Fisheries and Aquaculture.* Rome: FAO Fisheries Department, 2010. http://www.fao.org/docrep/013/i1820e/i1820e01.pdf.

Farber, Daniel. "BP Blowout and the Social and Environmental Erosion of the Louisiana Coast." *Minnesota Journal of Law, Science & Technology* 13 (2012): 37.

Farber, Daniel. "Issues of Scale in Climate Governance." In *The Oxford Handbook of Climate Change and Society*, edited by John S. Dryzek, Richard B. Norgaard, and David Schlosberg, 479–498. Oxford: Oxford University Press, 2011.

Federico, Carmela, and Jaime Cloud. "Kindergarten through Twelfth Grade Education: Fragmentary Progress in Equipping Students to Think and Act in a Challenging World." In *Agenda for a Sustainable America*, edited by John Dernbach, 109–127. Washington, DC: Environmental Law Institute Press, 2009.

Feinberg, Joel. "The Child's Right to an Open Future." In *Philosophy of Education*, edited by Randall Curren. 112–123. Oxford: Blackwell, 2007.

Feinstein, Noah. *Education for Sustainable Development in the United States of America: A Report Submitted to the International Alliance of Leading Education Institutes.* Madison: University of Wisconsin, 2009.

Feinstein, Noah, and Ginny Carlton. "Education for Sustainability in the K–12 Educational System of the United States." In *Schooling for Sustainable Development in Canada and the United States*, edited by Rosalyn McKeown and Victor Nolet, 37–49. New York: Springer, 2013.

Feinstein, Noah, and Kathryn L. Kirchgasler. "Sustainability in Science Education? How the Next Generation Standards Approach Sustainability, and Why It Matters." *Science Education* 99, no. 1 (2015): 121–144.

Finley, Moses. *Land and Credit in Ancient Athens, 500–200 B.C.* New Brunswick, NJ: Rutgers University Press, 1953.

Finley, Moses. *Politics in the Ancient World.* Cambridge: Cambridge University Press, 1983.

FitzPatrick, William J. "Valuing Nature Non-Instrumentally." *Journal of Value Inquiry* 38 (2004): 315–332.

Freeman, Richard B. *The Over-Educated American*. New York: Academic, 1976.

Frugoli, P. A., C. M. V. B. Almeida, F. Agostinho, B. F. Giannetti, and D. Huisingh. "Can Measures of Well-Being and Progress Help Societies to Achieve Sustainable Development?" *Journal of Cleaner Production* 90 (2015): 370–380.

Fullbrook, David. "Food Security in the Wider Mekong Region." In *The Water-Food-Energy Nexus in the Mekong Region*, edited by Alexander Smajgl and John Ward, 61–104. New York: Springer, 2013.

Fullinwider, Robert. "Moral Conventions and Moral Lessons." *Social Theory and Practice* 15, no. 3 (1989): 321–338.

Gale, Melanie, Merinda Edwards, Lou Wilson, and Alastair Greig. "The Boomerang Effect: A Case Study of the Murray-Darling Basin Plan." *Australian Journal of Public Administration* 73, no. 2 (2014): 153–163.

Galey, Sarah. "Education Politics and Policy: Emerging Institutions, Interests, and Ideas." *Policy Studies Journal: The Journal of the Policy Studies Organization* 43, no. 1 (2015): S12–S39.

Gardner, Howard, Mihaly Csikszentmihalyi, and William Damon. *Good Work*. New York: Basic Books, 2001.

Gardiner, Stephen M., Simon Caney, Dale Jamieson, and Henry Shue, eds. *Climate Ethics: Essential Readings*. Oxford: Oxford University Press, 2010.

Garnaut, Ross. *The Garnaut Climate Change Review*. Cambridge: Cambridge University Press, 2008.

Garvey, James. *The Ethics of Climate Change*. London: Continuum, 2008.

Gaspart, Frédéric, and Axel Gosseries. "Are Generational Savings Unjust?" *Politics, Philosophy & Economics* 6, no. 2 (2007): 193–217.

Gell-Mann, Murray. "Transformations of the Twenty-First Century: Transitions to Greater Sustainability." In *Global Sustainability: A Nobel Cause*, edited by Hans Joachim Schellnhuber, Mario Molina, Nicholas Stern, Veronika Huber, and Susanne Kadner, 1–9. Cambridge: Cambridge University Press, 2010.

Gert, Bernard. *Common Morality: Deciding What to Do*. New York: Oxford University Press, 2007.

Gifford, Robert. "The Dragons of Inaction: Psychological Barriers That Limit Climate Change Mitigation and Adaption." *American Psychologist* 66, no. 4 (2011): 290–302.

Gillis, Justin. "U.S. Climate Has Already Changed, Study Finds, Citing Heat and Floods." *New York Times*, May 7, 2014, A1, 13.

Gilovic, Thomas, Dale Griffin, and Daniel Kahneman. *Heuristics and Biases: The Psychology of Intuitive Judgment*. Cambridge: Cambridge University Press, 2002.

"Global Carbon Emissions Reach Record 10 Billion Tons, Threatening 2 Degree Target." *Science Daily*, December 6, 2011. http://www.sciencedaily.com/releases/2011/12/111204144648.htm.

Gold, Russell, and Ben Casselman. "On Doomed Rig's Last Day, a Divisive Change of Plan." *Wall Street Journal*, August 26, 2010. http://online.wsj.com/article/.

Goldman, Alvin. *Knowledge in a Social World*. New York: Oxford University Press, 1999.

Goodin, Robert. "Selling Environmental Indulgences." In *Climate Ethics: Essential Readings*, edited by Stephen M. Gardiner, Simon Caney, Dale Jamieson, and Henry Shue, 231–246. Oxford: Oxford University Press, 2010.

Goodman, Joan. "Student Authority: Antidote to Alienation." *Theory and Research in Education* 8, no. 3 (2010): 227–247.

Goodstein, David. *Out of Gas: The End of the Age of Oil*. New York: W. W. Norton & Company, 2004.

Graetz, Michael J. *The End of Energy: The Unmaking of America's Environment, Security, and Independence*. Cambridge, MA: MIT Press, 2013.

Grafton, R. Quentin, Gary Libecap, Samuel McGlennon, Clay Landry, and Bob O'Brien. "An Integrated Assessment of Water Markets: A Cross-Country Comparison." *Review of Environmental Economics and Policy* 5, no. 2 (2011): 219–239.

Grafton, R. Quentin, Jamie Pittock, Richard Davis, John Williams, Guobin Fu, Michele Warburton, Bradley Udall, Ronnie McKenzie, Xiubo Yu, Nhu Che, Daniel Connell, Qiang Jiang, Tom Kompas, Amanda Lynch, Richard Norris, Hugh Possingham, and John Quiggin. "Global Insights into Water Resources, Climate Change and Governance." *Nature Climate Change* 3, no. 4 (2013): 315–321.

Grafton, R. Quentin, Jamie Pittock, John Williams, Qiang Jiang, Hugh Possingham, and John Quiggin. "Water Planning and Hydro-Climatic Change in the Murray-Darling Basin, Australia." *Ambio* 43, no. 8 (2014): 1082–1092.

Graham, Bob, William K. Reilly, Frances Beinecke, Donald F. Boesch, Terry D. Garcia, Cherry A. Murray, and Fran Ulmer. *Deep Water: The Gulf Oil Disaster and the Future of Offshore Drilling*. Washington, DC: United States Publishing Office, 2011.

Green, Thomas F. *Predicting the Behavior of the Educational System*. Syracuse: Syracuse University Press, 1980.

Griffiths, Jacqui, and Rebecca Lambert. *Free Flow: Reaching Water Security through Cooperation*. Geneva: UNESCO, 2013.

Guidotti, Tee L. *Health and Sustainability: An Introduction*. Oxford: Oxford University Press, 2015.

Gupta, Joyeeta, Claudia Pahl-Wostl, and Ruben Zondervan. "Global Water Governance: A Multi-Level Challenge in the Anthropocene." *Current Opinion in Environmental Sustainability* 5, no. 6 (2013): 573–580.

Guston, David H. "Boundary Organizations in Environmental Policy and Science." *Science, Technology & Human Values* 26, no. 4 (2001): 399–408.

Gutentag, Edwin D., Frederick J. Heimes, Noel C. Krothe, Richard R. Luckey, and John B. Weeks. *Geohydrology of the High Plains Aquifer in Parts of Colorado, Kansas, Nebraska, New Mexico, Oklahoma, South Dakota, Texas, and Wyoming.* US Geological Survey Professional Paper 1400-B. Washington, DC: US Department of the Interior, 1984. http://pubs.usgs.gov/pp/1400b/report.pdf.

Gutmann, Amy, and Dennis Thompson. *Democracy and Disagreement.* Cambridge, MA: Harvard University Press, 1996.

Hall, Charles A. S., Jessica G. Lambert, and Stephen B. Balogh. "EROI of Different Fuels and the Implications for Society." *Energy Policy* 64 (2014): 141–152.

Hansen, Hal. "Rethinking Certification Theory and the Educational Development of the United States and Germany." *Research in Social Stratification and Mobility* 29 (2011): 31–55.

Hansen, James, et al. "Ice Melt, Sea Level Rise and Superstorms: Evidence from Paleoclimate Data, Climate Modeling, and Modern Observations that 2 °C Global Warming Could Be Dangerous." *Atmospheric Chemistry and Physics* 16 (March 22, 2016): 3761–3812. http://www.atmos-chem-phys.net/16/3761/2016/acp-16-3761-2016.html.

Hardin, Garrett. "The Tragedy of the Commons." *Science* 162 (1968): 1243–1248.

Head, Brian W. "Wicked Problems in Public Policy." *Public Policy* 3, no. 2 (2008): 101–118.

Head, Brian W., and John Alford. "Wicked Problems: Implications for Public Policy and Management." *Administration & Society* 47, no. 6 (2015): 711–739.

Heffernan, Margaret. *Willful Blindness: Why We Ignore the Obvious.* New York: Walker and Co., 2011.

Helevik, Ottar. "Beliefs, Attitudes, and Behavior towards the Environment." In *Realizing Rio in Norway: Evaluative Studies of Sustainable Development*, edited by William Laverty, Morton Nordskog, and Hilde A. Aakre, 7–19. Oslo: Program for Research and Documentation for a Sustainable Society, University of Oslo, 2002.

Herbertson, Kirk. "Xayaburi Dam: How Laos Violated the 1995 Mekong Agreement." January 13, 2013. https://www.internationalrivers.org/blogs/267/xayaburi-dam-how-laos-violated-the-1995-mekong-agreement.

Herman, Barbara. "Mutual Aid and Respect for Persons." In *Kant's Groundwork of the Metaphysics of Morals*, edited by Paul Guyer, 133–164. Lanham, MD: Rowman & Littlefield, 1998.

Herring, Stephanie C., Martin P. Hoerling, Thomas C. Peterson, and Peter A. Scott, eds. "Explaining Extreme Events of 2013 from a Climate Perspective." Supplement, *Bulletin of the American Meteorological Society* 95, no. 9 (September 2014), S1–S96. http://journals.ametsoc.org/doi/pdf/10.1175/1520-0477-95.9.S1.1.

Herzberg, Frederick. *Work and the Nature of Man*. New York: World Publishing, 1966.

Higgins, Peter, and Gordon Kirk. "Sustainability Education in Scotland: The Impact of National and International Initiatives on Teacher Education and Outdoor Education." *Journal of Geography in Higher Education* 30, no. 2 (2006): 313–326.

Ho, Ezra. "Unsustainable Development in the Mekong: The Price of Hydropower." *Consilience: The Journal of Sustainable Development* 12, no. 1 (2014): 63–76.

Hogan, David. "From Contest Mobility to Stratified Credentialing: Merit and Graded Schooling in Philadelphia, 1836–1920." *History of Education Review* 16 (1987): 21–42.

Holland, Alan. "Sustainability: Should We Start from Here?" In *Fairness and Futurity: Essays on Environmental Sustainability and Social Justice*, edited by Andrew Dobson, 46–68. Oxford: Oxford University Press, 1999.

Hopkins, Charles. "Education for Sustainable Development in Formal Education in Canada." In *Schooling for Sustainable Development in Canada and the United States*, edited by Rosalyn McKeown and Victor Nolet, 23–36. New York: Springer, 2013.

Horowitz, Joy. *The Poisoning of an American High School*. New York: Penguin, 2007.

Hou, Deyi, Jian Lou, and Abir Al-Tabbaa. "Shale Gas Can Be a Double-Edged Sword for Climate Change." *Nature Climate Change* 2 (June 2012): 385–387.

Hueston, Will, and Anni McLeod. "Overview of the Global Food System: Changes over Time/Space and Lessons for Future Food Safety." In *Improving Food Safety through a One Health Approach: Workshop Summary*, edited by Eileen R. Choffres, David A. Relman, Leigh Anne Olsen, Rebekah Hutton, and Alison Mack. Washington, DC: National Academies Press/Institute of Medicine, 2012. http://www.ncbi.nlm.nih.gov/books/NBK114491/.

Hughes, J. David. "A Reality Check on the Shale Revolution." *Nature* 494, no. 7437 (2013): 307–308.

Hylton, Wil S. "Broken Heartland: The Looming Collapse of Agriculture on the Great Plains." *Harper's Magazine* 325, no. 1946 (July 2012): 25–35.

International Energy Agency. "US Ethanol Production Plunges to Two-Year Low." IEA.org, August 13, 2012. https://www.iea.org/newsroomandevents/news/2012/august/us-ethanol-production-plunges-to-two-year-low.html.

International Energy Agency. "World Energy Outlook 2014 Factsheet." International Energy Agency, 2015. http://www.worldenergyoutlook.org/media/weowebsite/2014/141112_WEO_FactSheets.pdf.

IPCC. *Climate Change 2014: Synthesis Report*, edited by Core Writing Team, Rajendra K. Pachuri, and Leo Meyer. Geneva: IPCC, 2014. https://www.ipcc.ch/report/ar5/syr/.

IUCN. *Caring for the Earth: A Strategy for Sustainable Living*. Gland, Switzerland: IUCN, 1991.

Jacobs, Katherine L., Gregg M. Garfin, and M. Lenart. "More than Just Talk: Connecting Science and Decision Making." *Environment* 47, no. 9 (2005): 6–22.

Jamieson, Dale. *Reason in a Dark Time*. Oxford: Oxford University Press, 2014.

Jenkins, Willis. *The Future of Ethics: Sustainability, Social Justice, and Religious Creativity*. Washington, DC: Georgetown University Press, 2013.

Jickling, Bob. "Why I Don't Want My Children to Be Educated for Sustainable Development." *Journal of Environmental Education* 23, no. 4 (1992): 5–8.

Jickling, Bob, and Arjen E. J. Wals. "Globalization and Environmental Education: Looking beyond Sustainable Development." *Journal of Curriculum Studies* 40, no. 1 (2007): 1–21.

Johnston, Robyn M., Chu Thai Hoanh, Guillaume Lacombe, Andrew W. Noble, Vladimir Smakhtin, Dianne Suhardiman, Suan Pheng Kam, and Poh Sze Choo. *Rethinking Agriculture in the Greater Mekong Subregion: How to Sustainably Meet Food Needs, Enhance Ecosystem Services and Cope with Climate Change*. Colombo, Sri Lanka: International Water Management Institute, 2010.

Jones, Paula, David Selby, and Stephen Sterling. *Sustainability Education: Perspectives and Practice across Higher Education*. London: EarthScan, 2010.

Jorgenson, Andrew K., and Thomas Dietz. "Economic Growth Does Not Reduce the Ecological Intensity of Human Well-Being." *Sustainability Science* 10, no. 1 (2015): 149–156.

Kahn, Brian. "Drought Weakens the Amazon's Ability to Capture Carbon." *Climate Central*, March 9, 2015. http://www.climatecentral.org/news/drought-amazon-carbon-capture-18733.

Kahneman, Daniel, Paul Slovic, and Amos Tversky. *Judgment under Uncertainty*. Cambridge: Cambridge University Press, 1982.

Kahneman, Daniel, and Amos Tversky. *Choices, Values and Frames.* Cambridge: Cambridge University Press, 2000.

Kant, Immanuel. *Grounding for the Metaphysics of Morals.* Translated by James Ellington. Indianapolis: Hackett, 1981.

Kant, Immanuel. *The Metaphysics of Morals.* Translated by Mary Gregor. Cambridge: Cambridge University Press, 1991.

Kasser, Tim. *The High Price of Materialism.* Cambridge, MA: MIT Press, 2002.

Kasser, Tim, Steve Cohn, Allen Kanner, and Richard Ryan. "Some Costs of American Corporate Capitalism: A Psychological Exploration of Value and Goal Conflicts." *Psychological Inquiry* 18, no. 1 (2007): 1–22.

Kasser, Tim, and Allen Kanner. *Psychology and Consumer Culture: The Struggle for a Good Life in a Materialistic World.* 2nd ed. Washington, DC: American Psychological Association, 2013.

Kasser, Tim, and Richard M. Ryan. "Further Examining the American Dream: Differential Correlates of Intrinsic and Extrinsic Goals." *Personality and Social Psychology Bulletin* 22 (1996): 280–287.

Kasser, Tim, Richard M. Ryan, Charles E. Couchman, and Kennon M. Sheldon. "Materialistic Values: Their Causes and Consequences." In *Psychology and Consumer Culture: The Struggle for a Good Life in a Materialistic World*, edited by Tim Kasser and Allen Kanner, 11–28. Washington, DC: American Psychological Association, 2004.

Katz, Stanley. "Choosing Justice over Excellence." *Chronicle of Higher Education* 48, no. 35 (May 17, 2002): B7–B9. http://www.princeton.edu/~snkatz/papers/CHE_justice .html .

Katz, Stanley. "The Pathbreaking, Fractionalized, Uncertain World of Knowledge." *Chronicle of Higher Education* 49, no. 4 (September 20, 2002): B8. http://www .princeton.edu/~snkatz/papers/CHE_knowledge.html.

Keck, Margaret E., and Kathryn Sikkink. *Activists beyond Borders: Advocacy Networks in International Politics.* Ithaca, NY: Cornell University Press, 1998.

Kerr, Richard A. "Natural Gas from Shale Bursts onto the Scene." *Science* 328, no. 5986 (June 25, 2010): 1624–1626.

Kerr, Richard A. "Ocean Acidification Unprecedented, Unsettling." *Science* 328, no. 5985 (June 18, 2010): 1500–1501.

Keskinen, Marko, Joseph H. A. Guillaume, Mirja Kattelus, Miina Porkka, Timo A. Räsänen, and Olli Varis. "The Water-Energy-Food Nexus and the Transboundary Context: Insights from Large Asian Rivers," *Water* 8, no. 5 (2016). doi:10.3390/W8050193.

King, Megan F., Vivian F. Renó, and Evelyn M. L.M. Novo. "The Concept, Dimensions and Methods of Assessment of Human Well-Being within a Socioecological Context: A Literature Review." *Social Indicators Research* 116, no. 3 (2014): 681–698.

Kirby, Kris, and R. J. Herrnstein. "Preference Reversals Due to Myopic Discounting of Delayed Rewards." *Psychological Science* 6 (1995): 83–89.

Kitcher, Philip. "Public Knowledge and Its Discontents." *Theory and Research in Education* 9, no. 2 (2011): 103–124.

Kitcher, Philip. *Science in a Democratic Society*. Amherst, NY: Prometheus Books, 2011.

Kleinig, John. *Philosophical Issues in Education*. London: Croom Helm, 1982.

Komiyama, Hiroshi, and Kazuhiko Takeuchi. "Sustainability Science: Building a New Discipline." *Sustainability Science* 1, no. 1 (October 2006): 1–6. http://www.springerlink.com/content/214j253h82xh7342/fulltext.html.

Kraft, Jessica. "Running Dry." *Earth Island Journal* 28, no. 1 (Spring 2013): 47.

Kron, Wolfgang. "Increasing Weather Losses in Europe: What They Cost the Insurance Industry?" *CESifo Forum* 12, no. 2 (2011): 73–87.

Labaree, David. *How to Succeed in School without Really Learning: The Credentials Race in American Education*. New Haven, CT: Yale University Press, 1997.

Labaree, David. *The Making of an American High School: The Credentials Market and the Central High School of Philadelphia, 1838–1920*. New Haven, CT: Yale University Press, 1988.

Labaree, David. *Someone Has to Fail: The Zero-Sum Game of Public Schooling*. Cambridge, MA: Harvard University Press, 2010.

LaDue, Nicole D. "Help to Fight the Battle for Earth in US Schools." *Nature* 519, no. 7542 (2015): 131.

Lahsen, Myanna, Andrew Matthews, Michael R. Dove, Ben Orlove, Rajindra Puri, Jessica Barnes, Pamela McElwee, Frances Moore, Jessica O'Reilly, and Karina Yager. "Strategies for Changing the Intellectual Climate." *Nature Climate Change* 5 (May 2015): 391–392.

Lane, Melissa. *Eco-Republic: What the Ancients Can Teach Us about Ethics, Virtue, and Sustainable Living*. Princeton, NJ: Princeton University Press, 2012.

Larmer, Brook. "The Real Price of Gold." *National Geographic* 215, no. 1 (January 2009): 34–61.

Larsen, Christina. "Mekong Megadrought Erodes Food Security." *Science Magazine News*, April 6, 2016. http://www.sciencemag.org/news/2016/04/mekong-mega-drought-erodes-food-security.

Latham, Mark A. "BP Deepwater Horizon: A Cautionary Tale for CCS, Hydrofracking, Geoengineering and Other Emerging Technologies with Environmental and Human Health Risks." *William and Mary Environmental Law and Policy Review* 36, no. 1 (2011): 31–79.

Latham, Mark A. "Five Thousand Feet and Below: The Failure to Adequately Regulate Deepwater Oil Production Technology." *Boston College Environmental Affairs Law Review* 38, no. 2 (2011): 343–367. http://lawdigitalcommons.bc.edu/cgi/viewcontent.cgi?article=1692&context=ealr.

Layard, Richard. *Lessons from a New Science*. London: Penguin, 2005.

Layzer, Judith A. *Open for Business: Conservatives' Opposition to Environmental Regulation*. Cambridge, MA: MIT Press, 2014.

Lazarus, Eli. "Tracked Changes." *Nature* 529 (January 2016): 429.

Leiserowitz, Anthony. "American Risk Perceptions: Is Climate Change Dangerous?" *Risk Analysis* 25 (2005): 1433–1442.

Leiserowitz, Anthony, Edward Maibach, Connie Roser-Renouf, Geoff Feinberg, and Seth Rosenthal. *Climate Change in the American Mind: October 2015*. New Haven, CT: Yale Project on Climate Change Communication and George Mason University Center on Climate Change Communication, 2015. http://climatecommunication.yale.edu/wp-content/uploads/2015/11/Climate-Change-American-Mind-October-2015l.pdf.

Leiserowitz, Anthony, Edward Maibach, Connie Roser-Renouf, Geoff Feinberg, and Seth Rosenthal. *Global Warming and the U.S. Presidential Election, Spring 2016*. New Haven, CT: Yale Project on Climate Change Communication and George Mason University Center on Climate Change Communication, 2016. http://climatecommunication.yale.edu/wp-content/uploads/2016/05/2016_3_CCAM_Global-Warming-U.S.-Presidential-Election.pdf.

Leiserowitz, Anthony, Edward Maibach, Connie Roser-Renouf, and Jay Hmielowski. *Global Warming's Six Americas in March 2012 and November 2011*. New Haven, CT: Yale Project on Climate Change Communication and George Mason University Center for Climate Change Communication, 2012. http://environment.yale.edu/climate/files/Six-Americas-March-2012.pdf.

Lélé, Sharachchandra. "Sustainable Development: A Critical Review." In *Environment: An Interdisciplinary Anthology*, edited by Glenn Adelson, James Engell, Brent Ranalli, and K. P. Van Anglen, 144–152. New Haven, CT: Yale University Press, 2008.

Levin, Kelly, Benjamin Cashore, Steven Bernstein, and Graeme Auld. "Overcoming the Tragedy of Super Wicked Problems: Constraining Our Future Selves to Ameliorate Global Climate Change." *Policy Sciences* 45 (2) (2012): 123–152.

Lichtenberg, Judith. "Consuming because Others Consume." *Social Theory and Practice* 22, no. 3 (Fall 1996): 273–297.

Lipman, Matthew. *Thinking in Education.* 2nd ed. Cambridge: Cambridge University Press, 2003.

Lipschutz, Ronnie, and Corina Mckendry. "Social Movements and Global Civil Society." In *The Oxford Handbook of Climate Change and Society,* edited by John S. Dryzek, Richard R. Norgaard, and David Schlosberg, 369–383. Oxford: Oxford University Press, 2011.

Locke, John. *Second Treatise of Government.* Indianapolis: Hackett, 1989.

Lomborg, Bjørn. *The Skeptical Environmentalist.* Cambridge: Cambridge University Press, 2001.

MacKay, David. *Sustainable Energy—without the Hot Air.* Cambridge: UIT, 2009.

Mann, Horace. "Twelfth Annual Report." In *The Republic and the School: Horace Mann on the Education of Free Men,* edited by Lawrence Cremin, 79–112. New York: Teachers College Press, 1957.

Mark, Jason. "We Are All Louisianans." *Earth Island Journal* 25, no. 3 (Autumn 2010). http://www.earthisland.org/journal/index.php/eij/article/we_are_all_louisianans.

Markie, Peter. *A Professor's Duties: Ethical Issues in College Teaching.* Lanham, MD: Rowman & Littlefield, 1994.

Marsden, Terry, and Adrian Morley. "Current Food Questions and Their Scholarly Challenges." In *Sustainable Food Systems: Building a New Paradigm,* edited by Terry Marsden and Adrian Morley, 1–29. New York: Routledge, 2014.

Matthews, H. Damon, and Ken Caldeira. "Stabilizing Climate Requires Near-Zero Emissions." *Geophysical Research Letters* 35, no. 4 (2008): 1–5. doi:10.1029/2007GL032388.

Mays, Anthony, ed. *Disaster Management: Enabling Resilience.* Dordrecht, Netherlands: Springer, 2015.

Mazur, Laurie, ed. *A Pivotal Moment: Population, Justice and the Environmental Challenge.* Washington, DC: Island Press, 2010.

Mazur, Laurie, and Shira Saperstein. "Afterward: Work for Justice?" In *A Pivotal Moment: Population, Justice and the Environmental Challenge,* edited by Laurie Mazur, 393–396. Washington, DC: Island Press, 2010.

McAnany, Patricia, and Norman Yoffee, eds. *Questioning Collapse: Human Resilience, Ecological Vulnerability, and the Aftermath of Empire.* Cambridge: Cambridge University Press, 2010.

McArdle, Elaine. "What Happened to the Common Core?" *Harvard Ed. Magazine,* September 3, 2014. http://www.gse.harvard.edu/news/ed/14/09/what-happened -common-core.

McDonnell, Alexander B. "The Biggert-Waters Flood Insurance Reform Act of 2012: Temporarily Curtailed by the Homeowner Flood Insurance Act of 2014—A Respite to Forge an Enduring Correction to the National Flood Insurance Program Built on Virtuous Economic and Environmental Incentives." *Washington University Journal of Law and Policy* 49, no. 1 (2015): 235–268.

McIntosh, Roderick J., Joseph A. Tainter, and Susan Keech McIntosh, eds. *The Way the Wind Blows: Climate, History, and Human Action.* New York: Columbia University Press, 2000.

McKeown, Rosalyn. *Education for Sustainable Development Toolkit, Version 2.0.* 2007. http://www.esdtoolkit.org/.

McKeown, Rosalyn, and Victor Nolet, eds. *Schooling for Sustainable Development in Canada and the United States.* New York: Springer, 2013.

McKibben, Bill. "How Close to Catastrophe?" *New York Review of Books* 53 (November 16, 2006): 23–25.

McKibben, Bill. "The Pope and the Planet." *New York Review of Books* 62, no. 13 (August 13, 2015): 40–42.

McNeill, J. R. *Something New under the Sun: An Environmental History of the Twentieth-Century World.* New York: W. W. Norton, 2000.

McQuaid, John. "The Gulf of Mexico Oil Spill: An Accident Waiting to Happen." *Yale Environment* 360, May 10, 2010. http://e360.yale.edu/feature/the_gulf_of _mexico_oil_spill_an_accident_waiting_to_happen/2272/.

Meadows, Donella H. *Thinking in Systems: A Primer.* White River Junction, VT: Chelsea Green Publishing, 2008.

Mekong River Commission. "Agreement on the Cooperation for the Sustainable Development of the Mekong River Basin." Mekong River Commission, April 5, 1995. http://www.mrcmekong.org/assets/Publications/policies/agreement-Apr95.pdf.

Michaelis, Laurie. "Consumption Behavior and Narratives about the Good Life." In *Creating a Climate for Change: Communicating Climate Change and Facilitating Social Change,* edited by Susanne Moser and Lisa Dilling, 251–265. Cambridge: Cambridge University Press, 2007.

Millar, Andrew, and Douglas Navarick. "Self-Control and Choice in Humans." *Learning and Motivation* 15 (1984): 203–218.

Miller, David. "Political Philosophy for Earthlings." In *Political Theory: Methods and Approaches*, edited by David Leopold and Marc Stears, 29–48. Oxford: Oxford University Press, 2008.

Miller, Thaddeus R. *Reconstructing Sustainability Science*. London: Routledge, 2014.

Moellendorf, Darrell. *Cosmopolitan Justice*. Boulder, CO: Westview Press, 2002.

Molle, François, Tira Foran, and Mira Kakonen. *Contested Waterscapes in the Mekong Region: Hydropower, Livelihoods and Governance*. London: Earthscan, 2012.

Moore, Kathleen D., and Michael P. Nelson, eds. *Moral Ground: Ethical Action for a Planet in Peril*. San Antonio, TX: Trinity University Press, 2010.

Morgan, Ed. "Science in Sustainability: A Theoretical Framework for Understanding the Science-Policy Interface in Sustainable Water Resource Management." *International Journal of Sustainability Policy and Practice* 9 (2014): 37–54.

Morse, Stephen. "Developing Sustainability Indicators and Indices." *Sustainable Development* 23, no. 2 (2015): 84–95.

Moser, Suzanne C., and Maxwell T. Boykoff, eds. *Successful Adaptation to Climate Change*. London: Routledge, 2013.

Moser, Suzanne C., and Lisa Dilling, eds. *Creating a Climate for Change: Communicating Climate Change and Facilitating Social Change*. Cambridge: Cambridge University Press, 2007.

Murphy, David J. "The Implications of the Declining Energy Return on Investment of Oil Production." *Philosophical Transactions of the Royal Society of London A: Mathematical, Physical and Engineering Sciences* 372, no. 2006 (2014). doi:10.1098/rsta.2013.0126. http://rsta.royalsocietypublishing.org/content/372/2006/20130126.

Musiol, Erin, Nija Fountano, and Andreas Safakas. "Drought Planning in Practice." In *Planning and Drought*, edited by James C. Schwab, 43–74. Chicago: American Planning Association, 2013.

Myers, Ransom A., and Boris Worm. "Rapid Worldwide Depletion of Predatory Fish Communities." *Nature* 423 (May 15, 2003): 280–283.

Nagourney, Adam, Jack Healy, and Nelson D. Schwertz. "California Image vs. Dry Reality." *New York Times*, April 5, 2015, A1, 18–19.

Nash, Kate. "Towards Transnational Democratization?" In *Transnationalizing the Public Sphere*, edited by Kate Nash, 60–78. Cambridge: Polity Press, 2014.

National Research Council. *Climate Change Education in Formal Settings, K–14: A Workshop Summary*. Washington, DC: National Academies Press, 2012.

National Research Council. *Education for Life and Work: Developing Transferable Knowledge and Skills in the 21st Century*. Washington, DC: National Academies Press, 2012.

National Research Council. *A Framework for K–12 Science Education: Practices, Crosscutting Concepts, and Core Ideas*. Washington, DC: National Academies Press, 2012.

National Water Commission. *Australia's Water Blueprint: National Reform Assessment 2014*. Canberra: National Water Commission, 2014.

National Water Commission. *Water Markets in Australia: A Short History*. Canberra: National Water Commission, 2011.

Negin, Elliott. "Documenting Fossil Fuel Companies' Climate Deception." *Catalyst* 14 (Summer 2015): 8–11.

Nelson, Richard R. "Intellectualizing about the Moon-Ghetto Metaphor: A Study of the Current Malaise of Rational Analysis of Social Problems." *Policy Sciences* 5, no. 4 (1974): 375–414.

Newton, Lisa. *Ethics and Sustainability*. Upper Saddle River, NJ: Prentice-Hall, 2003.

NGSS Lead States. Next Generation Science Standards: For States, By States website. NGSS Lead States, 2013. http://www.nextgenscience.org/.

Niemiec, Christopher P., Martin F. Lynch, Maarten Vansteenkiste, Jessey Bernstein, Edward L. Deci, and Richard M. Ryan. "The Antecedents and Consequences of Autonomous Self-Regulation for College: A Self-Determination Theory Perspective on Socialization." *Journal of Adolescence* 29 (2006): 761–775.

Niemiec, Christopher P., Richard M. Ryan, and Edward L. Deci. "The Path Taken: Consequences of Attaining Intrinsic and Extrinsic Aspirations in Post-College Life." *Journal of Research in Personality* 43 (2009): 291–306.

Nolet, Victor. "Preparing Sustainability-Literate Teachers." *Teachers College Record* 111, no. 2 (2009): 409–442.

Nordhaus, William D. "A New Solution: The Climate Club." *New York Review of Books* 52, no. 10 (June 4, 2015): 36–39.

Norgaard, Kari Marie. *Living in Denial: Climate Change, Emotions, and Everyday Life*. Cambridge, MA: MIT Press, 2011.

Norton, Bryan. *Sustainability: A Philosophy of Adaptive Ecosystem Management*. Chicago: University of Chicago Press, 2005.

Norton, Bryan. *Sustainable Values, Sustainable Change*. Chicago: University of Chicago Press, 2015.

Nosich, Gerald. *Learning to Think Things Through: A Guide to Critical Thinking across the Curriculum*. 4th ed. Upper Saddle River, NJ: Pearson, 2011.

NRDC. "Groundbreaking Study Quantifies Health Costs of U.S. Climate Change-Related Disasters & Disease." National Resources Defense Council, November 8, 2011. https://www.nrdc.org/media/2011/111108-2.

NSW Government. N.d. "Algal Information." New South Wales Department of Primary Industries: Water. http://www.water.nsw.gov.au/Water-Management/Water-quality/Algal-information/Dangers-and-problems/Dangers-and-problems/default.aspx.

Nussbaum, Martha. *Frontiers of Justice*. Cambridge, MA: Harvard University Press, 2006.

Nussbaum, Martha. "Women's Education: A Global Challenge." *Signs* 29, no. 2 (2003): 325–355.

Ober, Josiah. *Democracy and Knowledge: Innovation and Learning in Classical Athens*. Princeton: Princeton University Press, 2008.

O'Connor, Robert E., Richard J. Bird, and Ann Fisher. "Risk Perceptions, General Environmental Beliefs, and Willingness to Address Climate Change." *Risk Analysis* 19 (1999): 461–471.

O'Donoghue, Ted, and Matthew Rabin. "Doing It Now or Later." *American Economic Review* 89 (1999): 103–124.

Oil Change International. "Fossil Fuel Subsidies: Overview." 2016. http://priceofoil.org/fossil-fuel-subsidies/.

O'Neill, Brian, F. Landin MacKellar, and Wolfgang Lutz. *Population and Climate Change*. Cambridge: Cambridge University Press, 2001.

O'Neill, Onora. "Consistency in Action." In *Kant's Groundwork of the Metaphysics of Morals*, edited by Paul Guyer, 103–131. Lanham, MD: Rowman & Littlefield, 1998.

O'Neill, Onora. "Constructivism in Rawls and Kant." In *The Cambridge Companion to Rawls*, edited by Samuel Freeman, 347–367. Cambridge: Cambridge University Press, 2003.

Oreskes, Naomi, and Erik Conway. *Merchants of Doubt*. New York: Bloomsbury Press, 2010.

Orr, David. "What Is Higher Education for Now?" In *State of the World 2010: Transforming Cultures, from Consumerism to Sustainability*, edited by Worldwatch Institute, 75–82. New York: W. W. Norton, 2010.

Ostrom, Elinor. "A General Framework for Analyzing Sustainability of Social-Ecological Systems." *Science* 325 (July 24, 2009): 419–422.

Ostrom, Elinor. "A Multi-Scale Approach to Coping with Climate Change and Other Collective Action Problems." *Solutions* 1, no. 2 (February 2010): 27–36.

Ostrom, Elinor. "Nested Externalities and Polycentric Institutions: Must We Wait for Global Solutions to Climate Change before Taking Actions at Other Scales?" *Economic Theory* 49, no. 2 (2012): 353–369.

Ostrom, Elinor. "Polycentric Systems for Coping with Collective Action and Global Environmental Change." *Global Environmental Change* 20, no. 4 (2010): 550–557.

Oxfam. *Education for Global Citizenship: A Guide for Schools.* Oxfam, 2015. http://www.oxfam.org.uk/education/global-citizenship/global-citizenship-guides.

Pearce, Fred. "UN Climate Report Is Cautious on Making Specific Predictions." *Environment 360*, March 24, 2014. http://e360.yale.edu/feature/un_climate_report_is_cautious_on_making_specific_predictions/2750/.

Pegram, Guy, Li Yuanyuan, Tom Le Quesne, Robert Speed, Li Jianqiang, and Shen Fuxin. *River Basin Planning Principles: Procedures and Approaches for Strategic Basin Planning.* Paris: UNESCO, 2013.

Pelletier, Luc, and Elizabeth Sharp. "Administrative Pressures and Teachers' Interpersonal Behavior." *Theory and Research in Education* 7, no. 2 (2009): 174–183.

Perlez, Jane, and Lowell Bergman. "Tangled Strands in Fight over Peru Gold Mines." *New York Times*, October 25, 2005. http://www.nytimes.com/2005/10/25/international/americas/25GOLD.html?th+&emc=th

Perlez, Jane, and Kirk Johnson. "Behind Gold's Glitter: Torn Lands and Pointed Questions." *New York Times*, October 24, 2005. http://www.nytimes.com/2005/10/24/international/24GOLD.html?th+&emc+th&pagewa

Peters, R. S. "Education as Initiation." In *Philosophy of Education: An Anthology*, edited by Randall Curren, 55–67. Oxford: Blackwell, 2007.

Peterson, H. C. "Sustainability: A Wicked Problem." In *Sustainable Animal Agriculture*, edited by Ermias Kebreab, 1–9. Wallingford, UK: CABI, 2013.

Piani, Adrian. "The Key Ingredients for Success of the Murray-Darling Basin Plan." *Global Water Forum.* 2013. http://www.globalwaterforum.org/2013/04/01/the-key-ingredients-for-successof-the-murray-darling-basin-plan/.

Pilz, Matthias. "Why *Abiturienten* Do an Apprenticeship before Going to University: The Role of 'Double Qualifications' in Germany." *Oxford Review of Education* 35, no. 2 (2009): 187–204.

Pimentel, David, and Tad Patzek. "Ethanol Production Using Corn, Switchgrass, and Wood; Biodiesel Production Using Soybean and Sunflower." *Natural Resources Research* 14, no. 1 (March 2005): 65–76.

Pittock, Jamie. "Devil's Bargain? Hydropower vs. Food Trade-Offs in the Mekong Basin." *World Rivers Review* 29, no. 4 (2007): 3, 14.

Plutarch. "Lycurgus." In vol. 1 of *Plutarch's Lives*, translated by B. Perrin, 52–80. Cambridge, MA: Harvard University Press, 1914.

Pollard, David, and Robert M. DeConto. "Contribution of Antarctica to Past and Future Sea-Level Rise." *Nature* 531, no. 7596 (March 31, 2016): 591–597.

Ponting, Clive. *A Green History of the World*. London: Penguin, 1991.

Porter, Eduardo. "How Renewable Energy Is Blowing Climate Change Efforts Off Course." *New York Times*, July 19, 2016. http://www.nytimes.com/2016/07/20/business/energy-environment/how-renewable-energy-is-blowing-climate-change-efforts-off-course.html?action=click&contentCollection=Politics&module=Related Coverage&region=EndOfArticle&pgtype=article.

Portney, Kent E. *Sustainability*. Cambridge, MA: MIT Press, 2015.

Portney, Kent E. *Taking Sustainable Cities Seriously: Economic Development, the Environment, and Quality of Life in American Cities*. 2nd ed. Cambridge, MA: MIT Press, 2013.

Potsdam Institute for Climate Impact Research and Climate Analysis. *Turn Down the Heat: Why a 4°C Warmer World Must Be Avoided*. Washington, DC: World Bank, 2013.

Powell, James. *The Inquisition of Climate Science*. New York: Columbia University Press, 2011.

Princen, Thomas, Michael Manites, and Ken Conca, eds. *Confronting Consumption*. Cambridge, MA: MIT Press, 2002.

Princen, Thomas, Jack P. Manno, and Pamela Martin. "Keep Them in the Ground: Ending the Fossil Fuel Era." In *State of the World 2013*, edited by Worldwatch Institute, 161–171. Washington, DC: Worldwatch Institute, 2013.

Pritchard, Michael. *Reasonable Children: Moral Education and Moral Learning*. Lawrence: University Press of Kansas, 1996.

Rachels, James, and Stuart Rachels. *The Elements of Moral Philosophy*. 5th ed. Boston: McGraw-Hill, 2007.

Raffaelle, Ryne, Wade Robison, and Evan Selinger, eds. *Sustainability Ethics: 5 Questions*. Copenhagen: Vince, Inc. Automatic Press, 2010.

Randhir, Timothy O. "Globalization Impacts on Local Commons: Multiscale Strategies for Socioeconomic and Ecological Resilience." *International Journal of the Commons* 10, no. 1 (2016): 387–404. https://www.thecommonsjournal.org/article/10.18352/ijc.517/.

Ravnborg, Helle Munk, and Maria del Pilar Guerrero. "Collective Action in Watershed Management—Experiences from the Andean Hillsides." *Agriculture and Human Values* 16, no. 3 (September 1999): 257–266.

Rawls, John. "Fairness to Goodness." In *John Rawls: Collected Papers*, edited by Samuel Freeman, 267–285. Cambridge, MA: Harvard University Press, 1999.

Rawls, John. *Justice as Fairness: A Restatement*. Cambridge, MA: Harvard University Press, 2001.

Rawls, John. *The Law of Peoples*. Cambridge, MA: Harvard University Press, 1999.

Rawls, John. *Political Liberalism*. New York: Columbia University Press, 1993.

Rawls, John. *A Theory of Justice*. Rev. ed. Cambridge, MA: Harvard University Press, 1999.

Redman, Charles. *Human Impact on Ancient Environments*. Tucson: University of Arizona Press, 1999.

Rees, William E., and Mathias Wackernagel. "The Shoe Fits, but the Footprint Is Larger than Earth." *PLoS Biology* 11, no. 11 (2013). doi:10.1371/journal.pbio .1001701.

Rickards, Lauren. "Power in Climate Change Research." *Nature Climate Change* 5 (May 2015): 392–393.

Rittel, Horst W. J., and Melvin M. Webber. "Dilemmas in a General Theory of Planning." *Policy Sciences* 4, no. 2 (1973): 155–169.

Rockström, Johan, Will Steffen, Kevin Noone, Åsa Persson, F. Stuart Chapin III, Eric F. Lambin, Timothy M. Lenton, Marten Scheffer, Carl Folke, Hans Joachim Schellnhuber, Björn Nykvist, Cynthia A. de Wit, Terry Hughes, Sander van der Leeuw, Henning Rodhe, Sverker Sörlin, Peter K. Snyder, Robert Costanza, Uno Svedin, Malin Falkenmark, Louise Karlberg, Robert W. Corell, Victoria J. Fabry, James Hansen, Brian Walker, Diana Liverman, Katherine Richardson, Paul Crutzen, and Jonathan A. Foley. "A Safe Operating Space for Humanity." *Nature* 461, no. 24 (September 2009): 472–475. http://www.nature.com/nature/journal/v461/n7263/full/461472a .html.

Rowe, Debra, Susan Jane Gentile, and Lilah Clevey. "The US Partnership for Education for Sustainable Development: Progress and Challenges Ahead." *Applied Environmental Education & Communication* 14, no. 2 (2015): 112–120.

Rumberger, Russell W. *Overeducation in the U.S. Labor Market*. New York: Praeger, 1981.

Ryan, Richard M. "Psychological Needs and the Facilitation of Integrative Processes." *Journal of Personality* 63 (1995): 397–427.

Ryan, Richard M., Randall Curren, and Edward L. Deci. "What Humans Need: Flourishing in Aristotelian Philosophy and Self-Determination Theory." In *The Best within Us: Positive Psychology Perspectives on Eudaimonia*, edited by Alan S. Waterman, 57–75. Washington, DC: American Psychological Association, 2013.

Ryan, Richard M., Veronika Huta, and Edward L. Deci. "Living Well: A Self-Determination Theory Perspective on Eudaimonia." *Journal of Happiness Studies* 9 (2008): 139–170.

Ryan, Richard M., Kennon M. Sheldon, Tim Kasser, and Edward L. Deci. "All Goals Are Not Created Equal: An Organismic Perspective on the Nature of Goals and Their Regulation." In *The Psychology of Action: Linking Cognition and Motivation to Behavior*, edited by Peter M. Gollwitzer and John A. Bargh, 7–26. New York: Guilford, 1996.

Ryan, Richard M., and Netta Weinstein. "Undermining Quality Teaching and Learning: A Self-Determination Theory Perspective on High-Stakes Testing." *Theory and Research in Education* 7, no. 2 (2009): 224–233.

Sachs, Jeffrey D. *The Age of Sustainable Development*. New York: Columbia University Press, 2015.

Schellnhuber, Hans Joachim, Mario Molina, Nicholas Stern, Veronika Huber, and Susanne Kadner. *Global Sustainability: A Nobel Cause*. Cambridge: Cambridge University Press, 2010.

Schlottmann, Christopher. "Educational Ethics and the DESD: Considering the Trade-Offs." *Theory and Research in Education* 6, no. 2 (2008): 207–219.

Scholte, Jan Aart, ed. *Building Global Democracy? Civil Society and Accountable Global Governance*. Cambridge: Cambridge University Press, 2011.

Schor, Juliet. *Born to Buy: The Commercialized Child and the New Consumer Culture*. New York: Scribner, 2004.

Schor, Juliet. *The Overspent American: Why We Want What We Don't Need*. New York: Harper, 1998.

Schor, Juliet. *The Overworked American: The Unexpected Decline of Leisure*. New York: Basic Books, 1992.

Schor, Juliet. *Plenitude: The New Economics of True Wealth*. New York: Penguin, 2010.

Schrader-Frechette, Kristin. *Taking Action, Saving Lives*. New York: Oxford University Press, 2007.

Schrag, Brian. "Moral Responsibility of Faculty and the Ethics of Faculty Governance." In *Ethics in Academia*, edited by S. K. Majumdar, Howard S. Pitkow, Lewis Penhall Bird, and E.W. Miller, 225–240. Easton: Pennsylvania Academy of Science, 2000.

Schrag, Daniel P. "Is Shale Gas Good for Climate Change?" *Daedalus* 141, no. 2 (2012): 72–80.

Scott, W. "Education and Sustainable Development: Challenges, Responsibilities, and Frames of Mind." *Trumpeter* 18, no. 1 (2002): 49.

Selby, David. "The Firm and Shaky Ground of Education for Sustainable Development." In *Green Frontiers: Environmental Educators Dancing Away from Mechanism*, edited by James Gray-Donald and David Selby, 59–75. Rotterdam: Sense Publishers, 2008.

Sen, Amartya. *Development as Freedom.* Oxford: Oxford University Press, 1999.

Sennett, Amy, Emma Chastain, Sarah Farrell, Tom Gole, Jasdeep Randhawa, and Chengyan Zhang. "Challenges and Responses in the Murray-Darling Basin." *Water Policy* 16, no. S1 (2014). doi:10.2166/wp.2014.006. http://wp.iwaponline.com/content/16/S1/117.

Shafer-Landau, Russ. *The Fundamentals of Ethics.* New York: Oxford University Press, 2010.

Shafer-Landau, Russ. *Whatever Happened to Good and Evil?* New York: Oxford University Press, 2004.

Sheldon, Kennon M., and Tim Kasser. "Coherence and Congruence: Two Aspects of Personality Integration." *Journal of Personality and Social Psychology* 68, no. 3 (1995): 531–543.

Sherren, Kate. "A History of the Future of Higher Education for Sustainable Development." *Environmental Education Research* 14, no. 3 (2008): 238–256.

Shu, Xiaoling, and Margaret Mooney Marini. "Coming of Age in Changing Times: Occupational Aspirations of American Youth, 1966–1980." *Research in Social Stratification and Mobility* 26, no. 1 (2008): 29–55.

Siegel, Harvey. *Educating Reason.* New York: Routledge, 1988.

Simmons, A. John. "Ideal and Non-Ideal Theory." *Philosophy & Public Affairs* 38, no. 1 (2010): 5–36.

Slovic, Peter. *The Perception of Risk.* London: Earthscan, 2000.

Smith, Adam. *Wealth of Nations.* Edited by R. H. Campbell and Andrew S. Skinner. Oxford: Clarendon Press, 1976.

Smith, Christian. *Lost in Transition: The Dark Side of Emerging Adulthood.* New York: Oxford University Press, 2011.

Smith, Kim, Peter Adriance, Alex Mueller, Rosalyn McKeown, Victor Nolet, Debra Rowe, and Madison Vorva. *The Status of Education for Sustainable Development (ESD) in the United States: A 2015 Report to the US Department of State.* International Society of Sustainability Professionals, December 2015. https://www.sustainabilityprofessionals.org/sites/default/files/ESD%20in%20the%20United%20States%20final.pdf.

Smith, Kim, Debra Rowe, Peter Adriance, Rosalyn McKeown, Victor Nolet, and Madison Vorva. "UNESCO Roadmap for Implementing the Global Action Programme on Education for Sustainable Development: Implementation Recommendations for the United States of America." US Delegation to the UNESCO World Conference on ESD, 2015. http://gpsen.org/wp-content/uploads/2016/05/GAP-Roadmap-Recommendations-Final.pdf.

Smyth, John C. *Are Educators Ready for the Next Earth Summit?* London: Stakeholder Forum, 2002.

Spearman, Mindy. "Sustainability Education." In *Educating about Social Issues in the 20th and 21st Centuries*, vol. 1, *A Critical Annotated Bibliography*, edited by Samuel Totten and Jon E. Pedersen, 251–269. Charlotte, NC: Information Age Publishers, 2012.

Speth, James. *America the Possible: Manifesto for a New Economy.* New Haven, CT: Yale University Press, 2012.

Speth, James. *The Bridge at the Edge of the World: Capitalism, the Environment, and Crossing from Crisis to Sustainability.* New Haven, CT: Yale University Press, 2008.

Speth, James. "The Limits of Growth." In *Moral Ground: Ethical Action for a Planet in Peril*, edited by Kathleen D. Moore and Michael P. Nelson, 3–8. San Antonio, TX: Trinity University Press, 2010.

Speth, James, and Peter Haas. *Global Environmental Governance.* Washington, DC: Island Press, 2006.

Steffen, Will, Katherine Richardson, Johan Rockström, Sarah E. Cornell, Ingo Fetzer, Elena M. Bennett, Reinette Biggs, Stephen R. Carpente, Wim de Vries, Cynthia A. de Wit, Carl Folke, Dieter Gerten, Jens Heinke, Georgina M. Mace, Linn M. Persson, Veerabhadran Ramanathan, Belinda Reyers, and Sverker Sörlin. "Planetary Boundaries: Guiding Human Development on a Changing Planet." *Science* 347, no. 6223 (2015). doi:10.1126/science.1259855.

Stemplowska, Zofia, and Adam Swift. "Ideal and Non-Ideal Theory." In *The Oxford Handbook of Political Philosophy*, edited by David Estlund, 373–392. New York: Oxford University Press, 2012.

Sterman, John D. "Communicating Climate Change Risks in a Skeptical World." *Climatic Change* 108, no. 4 (2011). doi:10.1007/s10584-011-0189-3. http://link.springer.com/article/10.1007%2Fs10584-011-0189-3.

Sterman, John D. "Learning from Evidence in a Complex World." *American Journal of Public Health* 96, no. 3 (2006): 505–514.

Stern, David. *The Environmental Kuznets Curve after 25 Years.* CCEP Working Paper 1514, Centre for Climate Economics and Policy, Crawford School of Public Policy,

Australian National University, December 2015. https://ccep.crawford.anu.edu.au/
sites/default/files/publication/ccep_crawford_anu_edu_au/2016-01/ccep1514_0.pdf.

Stern, Nicholas. *Stern Review on the Economics of Climate Change*. London: HM
Treasury, 2006.

Stern, Paul, Thomas Dietz, and Elinor Ostrom. "Research on the Commons: Lessons
for Environmental Resource Managers." *Environmental Practice* 4, no. 2 (June 2002):
61–64.

Stiglitz, Joseph E., Amartya Sen, and Jean-Paul Fitousi. *Mismeasuring Our Lives: Why
GDP Doesn't Add Up*. New York: The New Press, 2010.

Stout, Lynn. *Cultivating Conscience: How Good Laws Make Good People*. Princeton, NJ:
Princeton University Press, 2011.

Stout, Lynn. *The Shareholder Value Myth: How Putting Shareholders First Harms
Investors, Corporations, and the Public*. San Francisco: Berrett-Koehler Publishers,
2012.

Strauss, Valerie. "Revolt against High-Stakes Standardized Testing Growing—and So
Does Its Impact." *Washington Post*, March 19, 2015. http://www.washingtonpost
.com/blogs/answer-sheet/wp/2015/03/19/revolt-against-high-stakes-standardized
-testing-growing-and-so-does-its-impact/.

Strike, Kenneth. "The Ethics of Teaching." In *A Companion to the Philosophy of
Education*, edited by Randall Curren, 509–524. Oxford: Blackwell, 2003.

Strike, Kenneth. *Small Schools and Strong Communities: A Third Way of School Reform*.
New York: Teachers College Press, 2010.

Sunstein, Cass. *Worst-Case Scenarios*. Cambridge, MA: Harvard University Press,
2007.

Szasz, Andrew. "Is Green Consumption Part of the Solution?" In *The Oxford Handbook of Climate Change and Society*, edited by John S. Dryzek, Richard B. Norgaard,
and David Schlosberg, 594–608. Oxford: Oxford University Press, 2011.

Tainter, Joseph. *The Collapse of Complex Societies*. Cambridge: Cambridge University
Press, 1988.

Tainter, Joseph, T. F. H. Allen, and Thomas W. Hoekstra. "Energy Transformations
and Post-Normal Science." *Energy* 31 (2006): 44–58.

Tainter, Joseph, and Tadeusz Patzek. *Drilling Down: The Gulf Oil Debacle and Our
Energy Dilemma*. New York: Springer, 2012.

Tarrow, Sidney. *The New Transnational Activism*. Cambridge: Cambridge University
Press, 2005.

Termeer, Catrien. J. A. M., Art Dewulf, Gerard Breeman, and Sabina J. Stiller. "Governance Capabilities for Dealing Wisely with Wicked Problems." *Administration & Society* 47, no. 6 (2015): 680–710.

Teste, Jefferson W., Elisabeth M. Drake, Michael J. Driscoll, Michael W. Golay, and William A. Peters. *Sustainable Energy: Choosing among Options*. 2nd ed. Cambridge, MA: MIT Press, 2012.

Thompson, Andrea. "CO2 Nears Peak: Are We Permanently above 400 PPM?" *Climate Central*, May 16, 2016. http://www.climatecentral.org/news/co2-are-we -permanently-above-400-ppm-20351.

Thompson, Andrea. "Pope's Climate Encyclical: 4 Main Points." *Climate Central*, June 18, 2015. http://www.climatecentral.org/news/4-main-points-pope-climate -encyclical-19129.

"Timeline: Oil Spill in the Gulf." CNN.com. N.d. www.cnn.com/2010/US/05/03/ timeline.gulf.spill/index.html.

Tollefson, Jeff. "Antarctic Model Raises Prospect of Unstoppable Ice Collapse." *Nature* 531, no. 7596 (March 31, 2016): 562. http://www.nature.com/news/antarctic -model-raises-prospect-of-unstoppable-ice-collapse-1.19638.

Turner, Ralph. "Sponsored and Contest Mobility and the School System." *American Sociological Review* 25 (1960): 855–867.

Tyler, Tom. *Why People Obey the Law*. Princeton, NJ: Princeton University Press, 2006.

UNEP. *Climate Change Starter's Guidebook: An Issues Guide for Education Planners and Practitioners*. Paris: UNESCO/UNEP, 2011.

UNEP. *Global Environment Outlook 5*. Valletta, Malta: Progress Press, Ltd, 2012. http:// www.unep.org/geo/pdfs/geo5/GEO5_report_full_en.pdf.

UNEP. *Summary of the Sixth Global Environment Outlook, GEO-6, Regional Assessments: Key Findings and Policy Messages*. Nairobi: United Nations Environmental Programme, 2016. http://www.unep.org/publications/.

UNEP FI. *Universal Ownership: Why Environmental Externalities Matter to Institutional Investors*. Geneva: PRI Association and UN Environmental Programme Finance Initiative, 2011.

UNESCAP. *The Status of the Water-Food-Energy Nexus in Asia and the Pacific*. Bangkok: United Nations Economic and Social Commission for Asia and the Pacific, 2013. http://www.unescap.org/sites/default/files/Water-Food-Nexus%20Report.pdf.

UNESCO. Education for Sustainable Development: United Nations Decade (2005–2014) website. UNESCO, 2005. http://en.unesco.org/themes/education-sustainable -development.

UNESCO. *Guidelines and Recommendations for Reorienting Teacher Education to Address Sustainability.* Paris: UNESCO, 2005. http://unesdoc.unesco.org/images/0014/001433/143370e.pdf.

UNESCO. *Roadmap for Implementing the Global Action Programme on Education for Sustainable Development.* Paris: UNESCO, 2014. http://unesdoc.unesco.org/images/0023/002305/230514e.pdf.

UNESCO. *Shaping the Future We Want: UN Decade of Education for Sustainable Development (2005–2014) Final Report.* Paris: UNESCO, 2014. http://unesdoc.unesco.org/images/0023/002301/230171e.pdf.

UN Foundation. *The Millennium Ecosystem Assessment.* Geneva: UN Foundation, 2005. http://millenniumassessment.org/en/index.html.

UNFPA. *Programme of Action: Adopted at the International Conference on Population and Development, Cairo, 5–13 September 1994.* Geneva: United Nations Population Fund, 2004.

Union of Concerned Scientists. *The Climate Deception Dossiers.* Cambridge, MA: UCS, 2015. http://www.ucsusa.org/decadesofdeception.

Union of Concerned Scientists. *Smoke, Mirrors, and Hot Air: How ExxonMobil Uses Big Tobacco's Tactics to Manufacture Uncertainty on Climate Science.* Cambridge, MA: UCS, 2007.

Urbina, Ian, and Justin Gillis. "Workers on Oil Rig Recall a Terrible Night of Blasts." *New York Times,* May 7, 2010. http://www.nytimes.com/2010/05/08/us/08rig.html?pagewanted=all.

USDA Forest Service, Pacific Northwest Research Station. "The Healing Effects of Forests." *Science Daily,* July 26, 2010. http://www.sciencedaily.com/releases/2010/07/100723161221.htm

van den Bergh, Jeroen C., and Fabio Grazi. "Reply to the First Systematic Response by the Global Footprint Network to Criticism: A Real Debate Finally?" *Ecological Indicators* 58 (2015): 458–463.

Vansteenkiste, Maarten, Bart Neyrinck, Christopher Niemiec, Bart Soerens, Hans De Witte, and Anja Van den Broeck. "On the Relations among Work Value Orientations, Psychological Need Satisfaction and Job Outcomes: A Self-Determination Theory Approach." *Journal of Occupational and Organizational Psychology* 80 (2007): 251–277.

Vansteenkiste, Maarten, Bart Soenens, Joke Verstuyf, and Willy Lens. "'What Is the Usefulness of Your Schoolwork?' The Differential Effects of Intrinsic and Extrinsic Goal Framing on Optimal Learning." *Theory and Research in Education* 7, no. 2 (2009): 155–163.

Varghese, Shiney. "Water Governance in the 21st Century: Lessons from Water Trading in the U.S. and Australia". IATP, 2013. http://www.iatp.org/documents/water-governance-in-the-21st-century.

Venkatasubramanian, Venkat. "Systemic Failures: Challenges and Opportunities in Risk Management in Complex Systems." *AIChE Journal* 57, no. 1 (2011): 2–9.

Wackernagel, Mathias, and William Rees. *Our Ecological Footprint*. Gabriola Island, BC: New Society, 1996.

Wagner, Gernot, and Martin L. Weitzman. *Climate Shock: The Economic Consequences of a Hotter Planet*. Princeton, NJ: Princeton University Press, 2015.

Wallis, Philip J., and Raymond Ison. "Appreciating Institutional Complexity in Water Governance Dynamics: A Case from the Murray-Darling Basin, Australia." *Water Resources Management* 25, no. 15 (2011): 4081–4097.

Wassel, Raymond. "Lessons from the Macondo Well Blowout in the Gulf of Mexico." *Bridge* 44, no. 3 (2014): 46–53.

WCED (World Commission on Environment and Development). *Our Common Future*. Geneva: United Nations, 1987.

Weinstein, Netta, and Richard M. Ryan. "When Helping Helps: Autonomous Motivation for Prosocial Behavior and Its Influence on Well-Being for the Helper and Recipient." *Journal of Personality and Social Psychology* 98 (2010): 222–244.

Weissmann, Jordan. "America's Most Obvious Tax Reform: Kill the Oil and Gas Subsidies," *Atlantic*, March 19, 2013. http://www.theatlantic.com/business/archive/2013/03/americas-most-obvious-tax-reform-idea-kill-the-oil-and-gas-subsidies/274121/.

Wentworth Group of Concerned Scientists. "Statement on the Future of Australia's Water Reform." Wentworth Group of Concerned Scientists, October 10, 2014. http://wentworthgroup.org/2014/10/statement-on-the-future-of-australias-water-reform/2014/.

Weston, Burns H., and David Bollier. *Green Governance: Ecological Survival, Human Rights, and the Law of the Commons*. Cambridge: Cambridge University Press, 2014.

White, G. Edward. *Tort Law in America: An Intellectual History*. New York: Oxford University Press, 1980.

White House Press Office. "White House Report: The Every Child Succeeds Act." White House, Office of the Press Secretary, December 10, 2015. https://www.whitehouse.gov/the-press-office/2015/12/10/white-house-report-every-student-succeeds-act.

WHO (World Health Organization). *Climate and Health*. Geneva: WHO Media Centre, 2007.

WHO (World Health Organization). *Climate Change and Health, Fact Sheet No. 266, Revised August 2014*. Geneva: WHO Media Centre, 2014.

Wiedmann, Thomas, and John Barret. "A Review of the Ecological Footprint Indicator—Perceptions and Methods." *Sustainability* 2 (2010): 1645–1693.

Williams, Michael. *Deforesting the Earth: From Prehistory to Global Crisis (An Abridgment)*. Chicago: University of Chicago Press, 2006.

Wolf, Alison. *Does Education Matter? Myths about Education and Economic Growth*. London: Penguin, 2002.

Wolf, Susan. *Meaning in Life and Why It Matters*. Princeton, NJ: Princeton University Press, 2010.

World Wildlife Fund. "2012 Weather Extremes: Year-to-Date Review." December 6, 2012. http://www.wwfblogs.org/climate/sites/default/files/2012-Weather-Extremes-Fact-Sheet-6-dec-2012-final.pdf.

World Wildlife Fund. *Living Planet Report 2012: Biodiversity, Biocapacity and Better Choices*. Gland, Switzerland: WWF International, 2012. http://worldwildlife.org/publications/living-planet-report-2012-biodiversity-biocapacity-and-better-choices.

World Wildlife Fund. *Living Planet Report 2014: Species and Spaces, People and Places*. Gland, Switzerland: WWF International, 2014. http://wwf.panda.org/about_our_earth/all_publications/living_planet_report/.

World Wildlife Fund. *Living Blue Planet Report 2015*. Gland, Switzerland: WWF International, 2015. http://www.worldwildlife.org/publications/living-blue-planet-report-2015.

Worldwatch Institute. *State of the World 2013: Is Sustainability Still Possible?* Washington, DC: Island Press, 2013.

Worldwatch Institute. *State of the World 2015: Hidden Threats to Sustainability*. Washington, DC: Island Press, 2015.

Worldwatch Institute. *State of the World: Transforming Cultures from Consumerism to Sustainability*. New York: W. W. Norton & Company, 2010.

Worldwatch Institute. *Vital Signs 2012*. Washington, DC: Island Press, 2012.

Worldwatch Institute. *Vital Signs, Vol 22: The Trends That Are Shaping Our Future*. Washington, DC: Worldwatch Institute, 2015.

Worm, Boris, Edward B. Barbier, Nicola Beaumont, J. Emmett Duffy, Carl Folke, Benjamin S. Halpern, Jeremy B. C. Jackson, Heike K. Lotze, Fiorenza Micheli, Stephen R. Palumbi, Enric Sala, Kimberley A. Selkoe, John J. Stachowicz, and Reg Watson. "Impacts of Biodiversity Loss on Ocean Ecosystem Services." *Science* 3, no. 5800 (November 2006): 787–790.

Wright, Ronald. *A Short History of Progress*. Toronto: House of Anansi Press, 2004.

Wysession, Michael E. "Implications for Earth and Space in New K–12 Science Standards." *Eos, Transactions, American Geophysical Union* 93, no. 46 (2012): 465–466.

Wysession, Michael E. "The 'Next Generation Science Standards' and the Earth and Space Sciences." *Science and Children* 50, no. 8 (April 2013): 17–23. http://eric.ed.gov/?id=EJ1020542.

Zack, Naomi. *Ethics for Disaster*. Lanham, MD: Rowman & Littlefield, 2009.

Zellner, Moira, and Scott D. Campbell. "Planning for Deep-Rooted Problems: What Can We Learn from Aligning Complex Systems?" *Planning Theory & Practice* 16, no. 4 (2015): 457–478.

Index